SCIENCE AS A
CULTURAL EXPRESSION

SCIENCE AS A CULTURAL EXPRESSION

DAVID C. PEASLEE (ED.)

Nova Science Publishers, Inc.
Commack, New York

Editorial Production: Susan Boriotti
Office Manager: Annette Hellinger
Graphics: Frank Grucci and John T'Lustachowski
Information Editor: Tatiana Shohov
Book Production: Donna Dennis, Patrick Davin, Christine Mathosian
  and Tammy Sauter
Circulation: Maryanne Schmidt
Marketing/Sales: Cathy DeGregory

Library of Congress Cataloging-in-Publication Data

Peaslee, David C.
    Science as a Cultural Expression.

    ISBN 1-56072-589-3

Copyright © 1998 by Nova Science Publishers, Inc.
    6080 Jericho Turnpike, Suite 207
    Commack, New York 11725
    Tele. 516-499-3103   Fax 516-499-3146
    e-mail: Novascience@earthlink.net
    e-mail: Novasci1@aol.com
    Web Site: http://www.nexusworld.com/nova

Printed in the United States of America

Contents:

FOREWORD

The essays presented here originated as contributions, generally lectures, to a "Science Precept for Alumni Non-scientists" sponsored by the Princeton Class of 1943 on June 13-17, 1994. The aim of the precept was to expound the unity of science with general culture – in our times a relation of reciprocal support as never before.

Accordingly, the discussion sessions were organized into five groups: (1) initial learning processes underlying all the aspects of human culture; (2) trial and error as a source of human knowledge; (3) relation of science to the surrounding society; (4) cultural aspects external to science; and (5) distinctive features of science.

In the following the first five articles are taken one each from the categories in the sequence listed above. The last article is an attempt to draw explicit parallels between science and humanities.

The contributors to this volume follow in the order of their appearance:

Prof. Susan Sugarman, Psychology Department, Princeton University

Dr. Annmarie Sabb, Wyeth-Ayerst Research, Princeton, New Jersey

Prof. Bradley Dickinson, Electrical Engineering Department, Princeton University

Mr. Thomas P. Cook, Princeton Class of 1932

Prof. Frank Westheimer, Chemistry Department, Harvard University

Prof. David Peaslee, Physics Department, University of Maryland

The organizer wishes to thank these authors for their generous donations of time and effort. The precept seemed to share a common property of experiments in science; yielding upon completion a battery of hints for further experiments.

D.C. Peaslee
Class of 1943

WHO KNOWS THE WORLD? PARADOXES IN INFANTS' AND ADULTS' CONCEPTIONS OF REALITY

SUSAN SUGARMAN

PROLOGUE

This lecture series concerns the method of discovery in science. My contribution to it was supposed to have been a discussion of possible bases of the scientific method in the activities of infants and children. I would have argued as follows. Infants' and children's lives are all about observation and experiment. The older children become, the more subtle become their observations and the more complex their experiments. We retain these capacities as we grow into adults. Sadly, and probably to our detriment, we lose some of the naiveté of children. It has been said that the best science, and our best art, perhaps our greatest humanity, revive some of that naiveté[1] We do best when we assume, as children naturally do, that we do *not* know the answer, and hence must observe and experiment.

Rather than dwell on these parallels, I will apply them. I will adopt a self-consciously naive perspective in attempting to address a basic question:

How do children and adults perceive reality, the very object that scientists take for granted when they apply their tools of observation

---

[1]See, for instance, H. Gardner (1993), *Creating minds: An anatomy of creativity seen through the lives of Freud, Einstein, Picasso, Stravinsky, Eliot, Graham, and Ghandi* (New York: Basic), and references cited there.

and experiment? This turns out to be a complicated question, given some simple observations of children and adults.

I will approach the question through a combination of observation and reasoning. I will mention a few observations that bear on the question. These observations either are self evident or emerge readily from the research literature. In one way or another, they pose a conceptual puzzle. My object will be to sort through the puzzles. Scholarly inquiry, including scientific inquiry, involves the experimentation not only with physical entities, but with ideas. We call this the thought experiment, or the *gedankenexperiment*. I will use that process here.

I will center my discussion around two apparent paradoxes, one involving children (babies) and one involving adults. I begin with the one involving children.

BABIES' CONSTRUCTION OF REALITY

We adults experience a world of objects (substance), space, time, and causality. It is a world that we know as existing external to and independently of ourselves. It is our ordinary world of tables, chairs, people, food, cars, and so on. We interact with it and talk about it as though it were whole, solid, and external.

Skeptics may wonder how we can know these things: How *do* I know that the door in the back of the room is external to me and not a perturbation arising from the process of my looking? The ordinary person does not wonder about this. He or she thinks of the door as that thing out there that you can open and walk through. And he or she knows that there is another door up the stairs and around the corner, out of which he or she may exit – even though that door is not visible right now.

No less a writer than Kant[2] supposed that we must make some basic assumptions about the world *a priori*, just to get our tables, chairs, and people (etc.) going. People who have studied babies, for example the late Swiss psychologist Jean Piaget, have suggested

---

[2]I. Kant (1983). *Critique of pure reason* (N.K. Smith, translator). London: MacMillan (original German published 1787).

otherwise. We form the basic constituents of reality gradually through our experience. Only gradually and through their interactions with things do babies come to conceive, for example, a world that is external to them, that endures through space and time.[3]

If you watch babies, Piaget's idea seems to have some plausibility. Babies don't talk. So they can't, as we might do, say something that would indicate that they assume that an external, solid world exists out there. They do, however, behave. But *how* do they behave? They can't, as we might, pick themselves up and walk through the door to retrieve the wallet they left on the table outside. They can't walk, and before 7 or 8 months they can't crawl. Hence, they are limited in their capacity to show us that they believe that something exists out there beyond the room, even if they believe it does.

It turns out, moreover, that if you interest babies in an object, say, your keychain, and then remove it, they may (if they are very young) show no further interest in it. Slightly later (around 5 months), they may cry if you remove the keychain, but they won't pursue it, even if they have the physical capacity to do so. Suppose you hid the keychain in your hand, for instance, or drop your handkerchief over it on the table. Babies of 5 months are physically capable of grasping your hand or lifting a handkerchief. But they won't necessarily implement these actions when you clasp the keychain or drop your handkerchief over it. Although they may be sad when something is taken away, they exhibit no evidence that they know that the object is retrievable.

---

[3] J. Piaget (1954). *The construction of reality in the child*. New York: Basic (original French published 1937). Piaget and other developmental psychologists have portrayed Kant as having supposed that our categories of reality are innate in the sense of being available from birth. Kant, however, disclaimed the idea that people have any cognition prior to experience. He held, rather, that a conceptual apparatus develops in response to experience; however, the nature of the apparatus is founded not upon experience but upon the structure of a finite mind. Thus, according to Kant, our categories are pregiven in the sense that their ultimate shape is predetermined. Experience, however, is necessary to provoke their formation, and on this account they develop. Piaget would object to this weaker sense of innateness, as well as to the stronger (nonKantian, Cartesian) one that the categories are formed at birth. Experience, he says, shapes the categories rather than forming only the occasion for their unfolding.

On the basis of observations such as these, Piaget and others concluded that babies do not perceive or conceive a world of solid objects existing independently of themselves. Merely by interacting with the more limited world whose existence they do grant, however, they progressively construct a more "permanent," autonomous world.[4]

Piaget's claims are controversial. More recent researchers believe that babies' world is more organized than Piaget thought. They base their claims on ever subtler experiments with babies, and delicate manipulations of the conditions of Piaget's original experiences.[5] I am going to disregard this line of work today and follow Piaget's story instead. On the one hand, I think the implications of this later work are ambiguous.[6] On the other hand, if Piaget's account has a problem, it is a conceptual and not an empirical one. It is therefore appropriate to scrutinize it conceptually, on its own terms. I continue with Piaget's account.[7]

Piaget has a precise conception of how babies' progressive construction of reality comes about. It happens in stages. In each stage, babies make discoveries that, according to Piaget's view, resemble the scientific process of discovery. I am going to give an example of one of the transitions that Piaget envisions occurs.

There is evidence that babies do not reliably coordinate the worlds of vision and touch (or other sensory modalities, e.g., vision and hearing, vision and sucking). They may see an object they like and get excited, but they will not reach for it. They may grasp, say, a rattle, and wave it, but will not necessarily bring it before their eyes to look at it. After roughly five months, these coordinations become routine. An object seen is immediately reached for, and grasped. An object touched is brought before the eyes and studied.

---

[4]*Ibid.*, Chapter 1.

[5] For a review, see, for example, P. Harris (1993), Object permanence in infancy. In A. Slater & G. Bremner (Eds.) *Infant Development*, Hillsdale, NJ: Erlbaum, 102-121.

[6] Regarding this ambiguity, see, for instance, S. Sugarman (1987). The priority of description in developmental psychology. *International Journal of Behavioral Development, 10*, 391-414.

[7] The argument excerpted here is drawn from S. Sugarman (1993), Piaget on the origins of mind: A problem in accounting for the development of mental capacities. In E. Dromi (Ed.), *Language and cognition: A developmental perspective*. Norwood,

Observing this sequence of events, Piaget inferred that babies may not initially conceive such a thing as an "object" that can be both seen and touched. There are only seen things and touched things, or more accurately, seeing and touching, where these activities are indistinct from (what we know to be) their objects. How, then, might babies get the idea that there exist *objects*, that exist independently of any mode of perceiving (e.g., seeing, touching), that can be both seen and touched?

Piaget suggested that a process of the following sort might occur. Babies initially infer that there are seen things and touched things (viz., seeing and touching). However, any time any one sensory-motor system is aroused, babies naturally try to bring to bear all of their systems. Hence, when a baby sees your keychain, her hand might start flopping around too, as a result of general excitation. It will come to pass sooner or later that the baby's hand will touch the object. Once the object is felt, it will also be grasped. But then, notice what has happened. All of a sudden there is this thing that is being both looked at and touched – even though the baby never planned it that way!

Now the baby is poised for a major discovery, says Piaget. For, insofar as the baby conceived the object as a looked-at thing, she will notice that it is also felt. Hence, the object can't be just part of the act of looking (some sort of visual perturbation), because it is touched too. As any capable scientist would do, the baby, according to Piaget, will accept the "disconforming" evidence and revise her hypothesis. There are *things* in the world that can be both seen and touched. And in fact, if you observe babies' behavior, you can see the gradual convergence of looking and reaching behaviors.

Piaget documented comparable moments of transition in intelligence throughout infancy. The end of the sequence, at about 18 months, is a world of fully constructed objects, space, time, and causality.

Piaget's account contains a circularity, however. According to Piaget, at a crucial moment, babies recognize that a thing they conceived of as a seen-thing only (a part of vision, or whatever) is

N.J.: Ablex, 18-31.

also touched, and felt. Hence, he envisions they conclude, only a single thing must be present that can be both seen and felt.

But *how* are babies to recognize that the seen-thing is also being felt? They must conceive the thing as being independent of the seeing, such that they can attribute it with the additional quality of being touchable. The very intellectual gain we are trying to explain, however, is how babies come to conceive the thing as independent of the seeing (or of any act that results in its perception). Hence, the account must end up assuming the attainment it is attempting to explain.

Unfortunately, comparable tautologies exist at every major point of transition in Piaget's theory. Babies' *behavior* is developing exactly as Piaget says: Previously unintegrated behaviors are becoming coordinated and ever more precisely adjusted to the tasks they execute. The interpretations about how the *mind* is developing do not hold up. The upshot is that we do not know what conception of reality babies have, or, more to our purpose here, how the adult conception comes about. As some have argued, our basic categories of reality could be innate. Or, Piaget could be right that they are built up over time. If they are, however, we are in the dark about exactly when, and exactly how, they are created.

I want to leave babies for a while, and turn to adults. I want to look at their behavior to see if we can gain any clarity on the question of what "reality" is to them and how it comes to be.

ADULT SKEPTICS?

As I said before, if you observe our everyday behavior and speech, you cannot but conclude that we conceive an automatically existing, external reality. There are tangible things in the world. They exist whether or not we are interacting with them at the moment. The real world just *is*, to us.

Except it isn't, quite. Sometimes we betray a skepticism or uncertainty, or at least we appear to do so. I became alerted to this circumstance in reading a lesser-known paper of Freud's, in which he recounts the following personal experience.[8]

When he was 48, Freud, a long-time admirer of antiquity, made his first trip to Athens. Upon reaching the top of the Acropolis, he had the sudden thought, "So all this really *does* exist, just as we learned at school!" An astute observer of human foibles, Freud noticed that this thought did not entirely make sense. He had never doubted that the Acropolis exists. Yet here he was affirming it as though he had doubted it.

The experience Freud had is common. Often people arrive at a place they have known about but have never seen. Their first thought upon their arrival is that it really does exist after all. The thought is as paradoxical in their case as it was in Freud's. They, too, have never doubted the existence of the place.

For the remainder of my talk I want to focus on this paradoxical experience of surprise. I will call it "Freud's thought," or "the thought," for convenience; I mean the general experience, not only Freud's own specific one.

Here is the question: If we take the existence of objects, both near and far, for granted, how do we come to express surprise at their existence when we see them? My pursuit of this question will lead me eventually to reflections on how we relate to reality generally.[9]

We arrive on the Acropolis, or in front of the Eiffel Tower, for the first time. We find ourselves surprised that they really exist after all. Why? Let us look at some common-sensical answers and assess how well they illuminate the "thought."

Perhaps we are simply awed by what we see. We may be in awe. But that is not what the "thought" expresses. It says, "So, it is *true*, after all, that all this really exists."

---

[8] S. Freud (1936). A disturbance of memory on the Acropolis: An open letter to Romain Rolland on the occasion of his seventieth birthday. *Pelican Freud Library, Vol. 11* (A. Richards, Vol. ed.). Harmondsmouth, Eng.: Penguin (1984), 443-456.

[9] This discussion is drawn from S. Sugarman (1998), *Freud on the Acropolis: Reflections on a paradoxical response to the real.* Boulder, CO.: Westview Press (in press). The book, which was completed three years after the present paper was delivered, evolved beyond the discussion in this section of the paper, which I have left essentially unchanged from its original version.

Ah, you say. The Acropolis is a phenomenal place. It is hard to believe it ever existed or that the civilization that created it ever evolved. Well and good. But as you travel to Athens, you don't wonder whether the physical remains that stand on the Acropolis are going to be there when you land. You're taking a trip to see them! Yet the thought affirms that all this, right here, exists after all.

But, you may think, seeing something is different from hearing or reading about it. So it is. Until you see it, you don't know entirely what it is like. You don't know how it feels. True enough. But now knowing what the thing is *like* is a different state from doubting whether it *exists*. The thought implies that you doubted that it exists.

Countless other interpretations are imaginable. You may be thinking of some already. Having studied the possibilities extensively, however, I can tell you that this simple and seemingly obvious thought has no straightforward explanation.

Compare the thought: "I think the tickets are in my back pocket." If I ask you why someone might have this thought, you can produce obvious, adequate answers. The person was wondering where her tickets were. She was musing that sometimes people arrive at the airport only to discover that they have forgotten their tickets. There just doesn't seem to be an equally natural and complete antecedent for what I am calling "Freud's thought."

Hence, Freud was right. The "thought" is an anomaly. Freud believed that every behavior, however, except for genuinely random behavior (if there is any), makes sense, in *some* context. If it does not appear to cohere within any common-sensical context, then it may make sense within a context of which we are not aware. That context would consist of some set of unconscious impulses, rooted somewhere in the person's history. Freud explained many apparently aberrant behaviors in this way, for instance, dreams, neurotic symptoms, and so-called "mistakes" of normal waking life, such as slips of the tongue.

Freud tried to interpret his own version of the "thought" in this way. He felt *guilty*, he inferred, for being on the Acropolis. His parents had been poor, and also uneducated. They would have had neither the means nor the interest to allow a trip to Athens.[10] We have

---

[10] Carl Shorske (1993. Freud's Egyptian dig. *The New York Review of Books*, May

an instinctive fear, Freud believed, of surpassing or criticizing our parents. Thus, he concludes, his thought about the Acropolis was a "displacement." It was a *defense* against the thought he really wanted to have: "Here I am, despite everything, on top of the Acropolis!" Afraid to acknowledge this thought, he exulted instead that the *Acropolis* exists!

As I have said, I think Freud is right that we must find something more than a common-sense context to explain the thought. And I think that he provides apt interpretations of numerous phenomena. But, after extensive study, I do not think that his account of the "thought" can explain the thought.

Suppose Freud was feeling guilty for being on the Acropolis, as he says. Suppose further, with Freud, that the thought acted in his case to camouflage the underlying thought, that *he* was on the Acropolis. If what he needed on the spot was a good defense, there are other ways in which he could have defended himself. He could have thought about something else: the weather, the alignment of columns on the Parthenon. Why choose specifically the exclamation that the Acropolis really exists? More to our purpose: Why does this same thought occur to other people, under the same circumstance of seeing for the first time something they have known about but never seen?[11]

Other facts about the thought are poorly accommodated by Freud's account. When we have the thought – "So all this (whatever) really *does* exist!" – we take particular *pleasure* in the observation and feel *release* through it. These are not the qualities of a defensive thought or behavior. When engaged in (conscious or unconscious) defensive behavior, we feel more constricted or suppressed.

27, *XLI* (10), 35-40.) adds the possibility that Freud, a Jew, felt guilty also about his preoccupation with a culture unrelated to Judaism.

[11] For completeness, I add the observation that people can have the thought when encountering quite mundane objects or events. All that matters is that one heard or read about the thing before, and now one is seeing it first hand. The thought does not, therefore, respond specifically to the grandeur of its object. Moreover, one can have the thought when one sees something *second* hand, after having seen it first hand. For instance, suppose that while out hiking you pass the remains of a Native-American village, and it is labelled as such. Some time later, you see the village marked on an area map. You think, "So it really *does* exist!" Hence, the thought is no simple instance of the adage that seeing is believing (e.g., B. Russell [1962]. *An inquiry into meaning and truth*. Baltimore: Penguin).

If neither common-sense nor an account modelled after Freudian psychodynamics can account for a behavior, then it is difficult to know where to turn for an explanation. That is part of what makes this behavior interesting. I continue next with some further attempts to explain it.

After studying the thought for a while, I had the idea that it was *childlike*. Only from the point of view of a very naive observer would it come as a surprise that the thing one has heard about really does exist, out there in the world. Additionally, the pleasurable character of the thought could be likened to a child's glee – in what is obvious to the big people.

This characterization of the thought as childlike solves the logical problem that the thought raises, that it appears to answer a doubt, when we have none. Insofar as we "speak" from the vantage point of a child when we have the thought, the pretext for a doubt is present.

For a long time, I believed that this was the way (or at least a way) to account for the thought. It is childlike, in both substance and mood. I gradually came to perceive inadequacies in this account, however. Why should an adult suddenly lapse into a child's point of view on the occasions that prompt the thought? Contrary to Freud's account, the lapse seems not to be serving any defensive purpose. It is not pathological in any way. We seem, rather, to be having fun. Perhaps then, we lapse to achieve pleasure that we would not otherwise achieve. In seeing our present experience with the simplicity and enthusiasm of a child, we refresh it.

But there are many childlike stances that we could adopt that would accomplish this rejuvenation. We could admire other simple aspects of the things we observe, for example, their shape or color. Why comment specifically on the existence of things, why affirm it, *when we had no doubt about it?* To achieve a complete explanation of the thought, we have to return to our original question about it: How do we come to allude to a doubt that (so far as we can tell) we never had? The idea that the thought is childlike does not entirely solve this problem.

After still further work, I have come to the conclusion that this thought/experience just *is* paradoxical. But, it is not the only instance in which our relation to reality is paradoxical, or at least obscure. Here is another instance.

Why do we like to see things we know about, in the first place? More to the point in the context of our present concern is the question: *Who* likes to see things he or she knows about? It is the person who believes fully in the existence of the thing. The person for whom it is important, say, to visit Auschwitz is not the skeptic or the person who would deny what happened there. It is the person who knows and understands fully what happened. That person does not seek "evidence" of what occurred, but an opportunity to connect with it. To stand on the soil upon which the Jews trod is to revive that past, to touch it and feel it. Nothing can replace the actual site – with or without standing ruins – in producing this effect.

But our interest in seeing the "real thing" does not stop with relics of the past, that may, through our contact with them, restore that past for us. We want to see the real version also of things unattached to any past associations.

This past winter, a spectacular, but harmless, flood occurred near my house, during one of the storms we had. I happened to be working at home that day, and saw the flood through my window. It occurred to me that if I had been away from home and had learned of the flood only afterwards, I would have regretted not having seen it. Why would I have regretted it?

The answer can't be that I wanted to verify that the flood really occurred. My regret at not having seen it would have been predicated exactly on the assumption that it had occurred.

The answer may be in part that I would want to have seen what the flood looked like. But that isn't all. Had an enterprising townsperson been filming the whole thing, and had I seen the film, I would still have been disappointed. Even if I could have been placed in a full-scale replica of the event, able to feel and smell it, as well as see it, I would still not have been satisfied. I would want to have witnessed the real thing.

Why? Why is a full-scale, multi-sensory replica no good? The question is surprisingly difficult to answer. The replica just isn't as good. We want direct, first-hand contact.

I think "Freud's thought" is related to this desire, perhaps need, for first-hand contact with things. One reason I think this is the following thought experiment. I suspect that if you asked the naive viewer why the replica of the flood would not be as good as the thing in itself, you might get an answer very like "Freud's thought." That person might say to you, "I would want to have seen it to be sure it really happened." As I have noted, the idea that we (think we) would want to have seen the flood to make sure it really happened is paradoxical. The desire to see it is predicated exactly on the belief that it occurred.

The circumstance surrounding the canonical version of Freud's thought is parallel. The canonical version of the thought occurs on occasions when we have attained the object of our desires. We think "So [all this] really *does* exist after all..." It is a precondition of this thought also that we have had absolutely no doubt about the matter, and that absence of doubt is the whole reason we are there.

But here as well, we have made first-hand contact with the outer world. We express this attainment oddly. We conflate the idea that we have made contact with the real-world manifestation of something with the idea that the thing exists in the world. And, we acknowledge a skepticism on this latter point that we never had.

Somehow this paradoxical assertion articulates our appreciation of first-hand contact, which when realized on some occasions leads to a kind of epiphany. Exactly what is in the epiphany remains obscure, and so, therefore, does the "thought" that expresses it.

BABIES, ADULTS AND REALITY

How do we, first as babies and then as adults, perceive reality? The answer is somewhat obscure. Babies' behavior is compatible with a view of the world that acknowledges no stable reality external to the self. Unfortunately, if we assume the existence of this view, as a "minimal" interpretation of the data, we then have difficulty imagining how babies ever revise the view. It is difficult to know on what basis they would ever infer that an external, enduring reality exists, no matter how precisely intercoordinated their behavior becomes. Hence, it is tempting to suppose that this idea is somehow a part of our endogenous mental make-up, as Kant envisioned.[2]

And yet, adults seem to betray an uncertainty about what exists, or at least a memory of impression of such uncertainty. They do this even though in all other respects they behave as Kant said: their basic world of tables and chairs is stable and enduring. Their actions suggest this, and so do their words.

It may be tempting to suppose that we start out with the inchoate idea of reality that Piaget attributes to babies. We then develop the coherent Kantian view, but retain memories of the earlier inchoateness. As I have said, it is difficult to envision how babies would overcome the inchoate view, and it is difficult to know what memory, if any, adults who have "Freud's thought" are resurrecting. These pieces of our natural epistemology must, I feel, remain obscure for the present. But they are no less intriguing for their elusiveness.

I return to the subject of scientists and babies, and ordinary people (adults):

Scientists are for more intents and purposes like ordinary adults. They assume the existence of a stable and enduring reality. They dissect and manipulate it to learn its further attributes. It is the philosopher who stands aside (or can) and asks how we can know that reality exists. The ordinary person is scientist, philosopher, and child. A stable, enduring reality is assumed, in day to day actions and thoughts. At the same time as we assume this reality, however, we also betray a questioning of it, reminiscent of the philosopher's ploy. But like a child, we are capable of rejoicing in the game.

Juxtaposed, these inclinations contradict one another. Somehow, however, we manage to navigate in the world, and to study it, without succumbing to unmitigated confusion. This must mean that our different perspectives coexist peacefully. As a result, our experience of the world, and of ourselves, must be all the richer.

REFERENCES

Freud, S. (1936). A disturbance of memory on the Acropolis: An open letter to Romain Rolland on the occasion of his seventieth birthday. *Pelican Freud Library, Vol. 11* (A Richards, Vol. Ed.). Harmondsmouth, Eng.: Penguin (1984), 443-456.

Gardner, H. (1993). *Creating minds: An anatomy of creativity seen through the lives of Freud, Einstein, Picasso, Stravinsky, Eliot, Graham and Ghandi*. New York: Basic.

Harris, P. (1993). Object permanence in infancy. In A. Slater & G. Bremner (Eds.), *Infant development*. Hillsdale, NJ: Erlbaum, 103-121.

Kant, I. (1983). *Critique of pure reason* (N.K. Smith, translator). London: MacMillan (original German published in 1787).

Piaget, J. (1954). *The construction of reality in the child*. New York: Basic. (Original French published in 1937).

Russell, B. (1962). *An inquiry into meaning and truth*. Baltimore: Penguin.

Shorske, C. (1993). Freud's Egyptian dig. *The New York Review of Books*, May 27, *XLI* (10), 35-40.

Sugarman, S. (1987). The priority of description in developmental psychology. *International Journal of Behavioral Development*, *10*, 391-414.

Sugarman, S. (1993). Piaget on the origins of mind: A problem in accounting for the development of mental capacities. In E. Dromi (Ed.), *Language and cognition: A developmental perspective*. Norwood, N.J.: Ablex, 18-31.

Sugarman, S. (1998). *Freud on the Acropolis: Reflections on a paradoxical response to the real*. Boulder, CO: Westview (in press).

STRATEGIES TOWARD THE DISCOVERY OF AN ALZHEIMER'S TREATMENT

*ANNMARIE L. SABB, *86*

What is Alzheimer's disease? Alzheimer's disease is a progressive degenerative brain disease which affects the elderly (1 in 4 over the age of 80). A consequence of this disease is that the size of the brain (Figure 1, normal human brain) is greatly reduced. This loss in size is not due to loss of water but to the death of nerve cells called neurons. One especially vulnerable area of the brain is the hippocampus. The amount of cell death in this area can be correlated with the degree of dementia in Alzheimer's disease. Figure 2 shows a normal brain which has been dissected in half to show the hippocampus. The hippocampus is just above the brainstem. In the diagram shown at the lower right of Figure 2, the hippocampus is marked with a capital H. Notice that there are projections leading from the hippocampus to the outer surface of the brain called the cortex (dashed lines). The area of the brain marked OC is the occipital cortex where the processing of visual images occurs. The front of the brain marked FC is the frontal cortex where associative thought is believed to take place. In Figure 3, the brain has been sliced in half giving a view from front to back. Notice that the hippocampus is small and that there is a part of it on each side of the brain. Again, notice that there are projections going from the hippocampus up to the cortex. I will now increase our magnification a little more. The area which appears in the box in Figure 3 has now been magnified (Figure

NORMAL HUMAN BRAIN

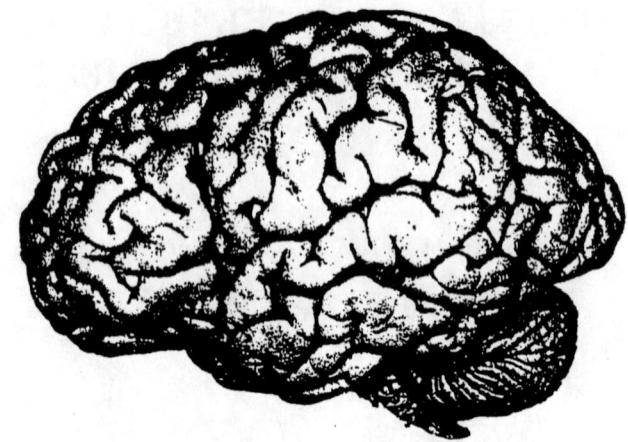

Figure 1

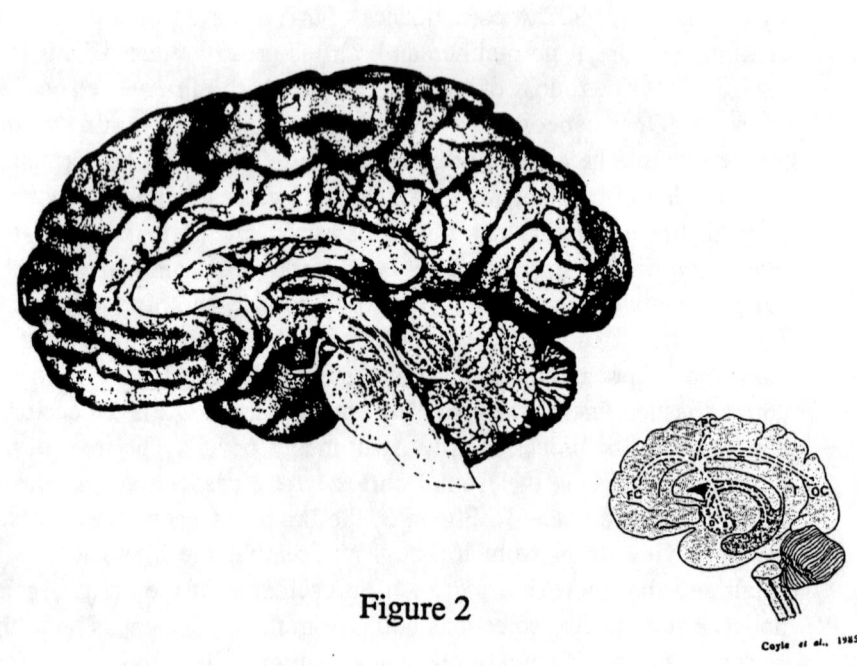

Figure 2

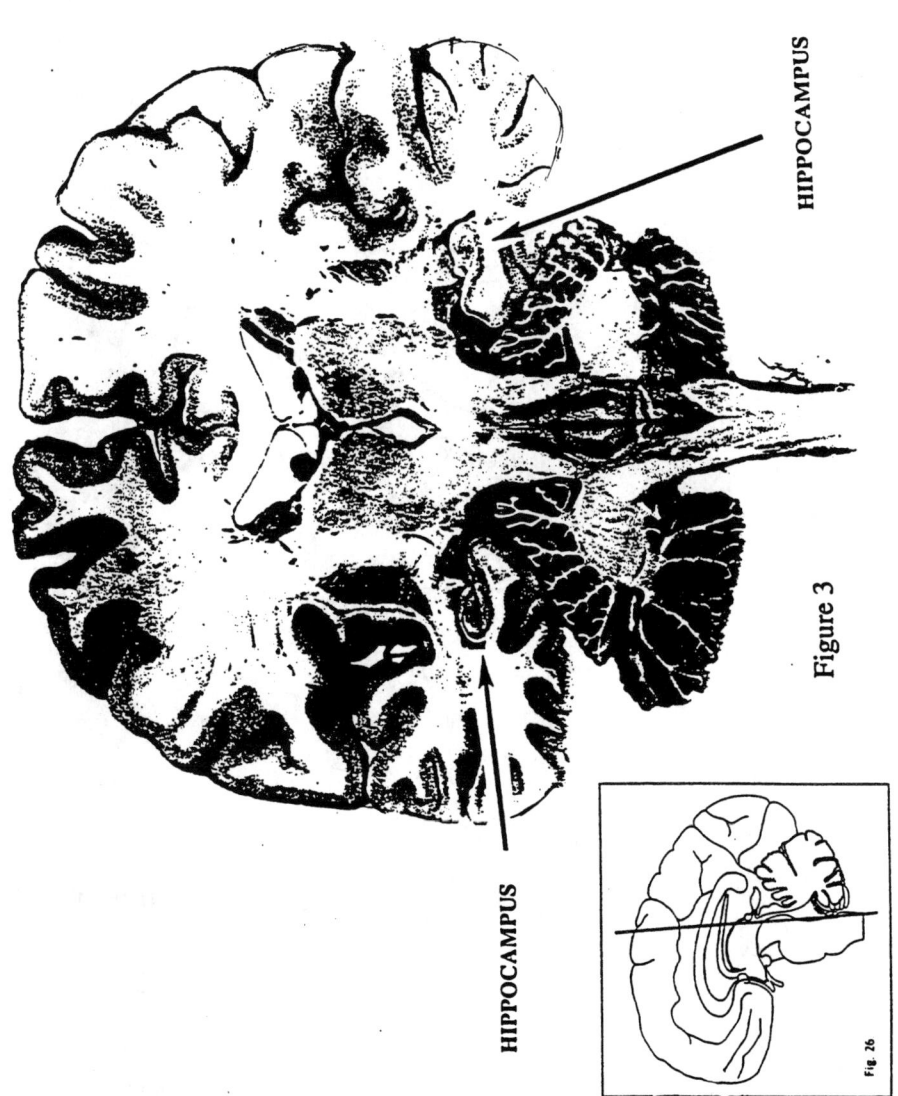

HIPPOCAMPUS

HIPPOCAMPUS

Figure 3

Fig. 26

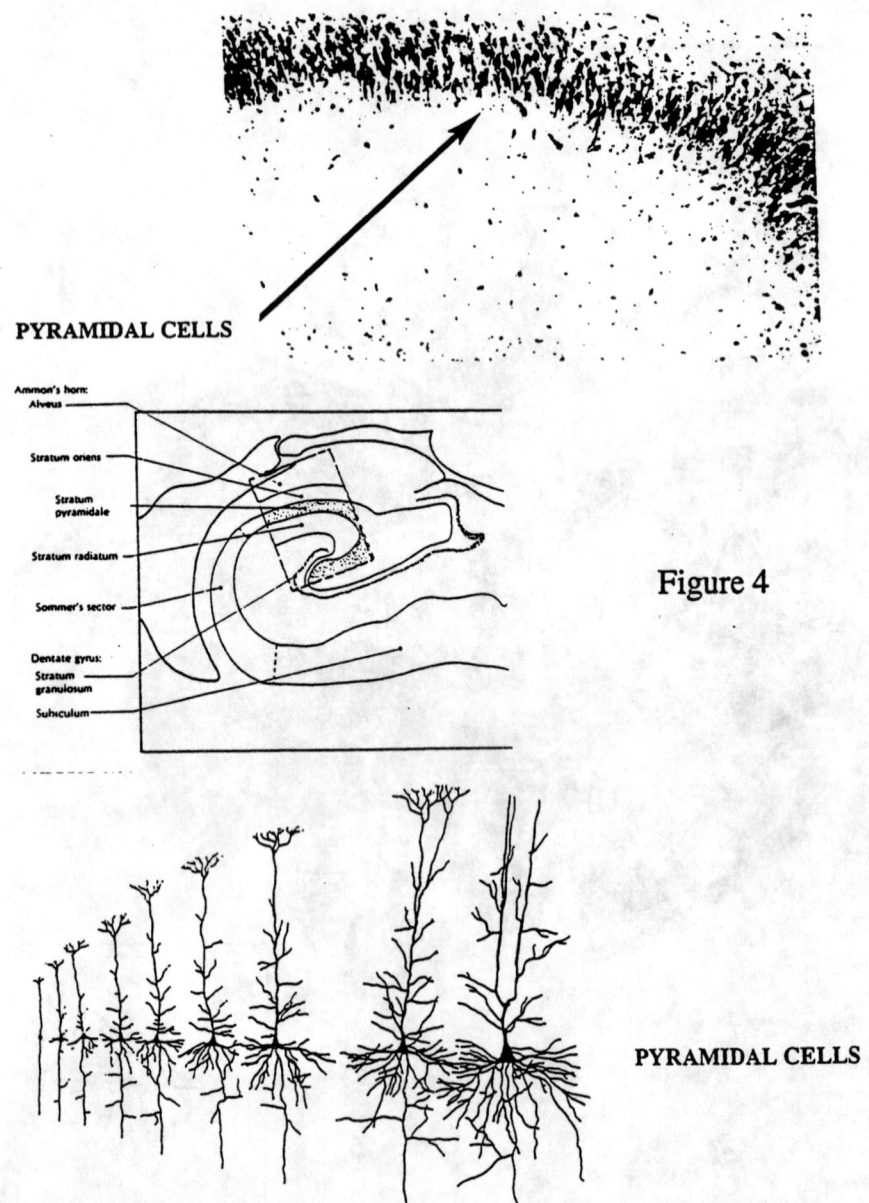

PYRAMIDAL CELLS

Figure 4

PYRAMIDAL CELLS

Figure 5 Growth and differentiation of the dendritic trees and axon collaterals of cortical pyramidal cells in the human, from fetus to adult. (Courtesy of P. Rakic)

Figure 6

ALZHEIMER'S DISEASE

SENILE PLAQUES AND NEUROFIBRILLARY TANGLES

NORMAL NEURONS

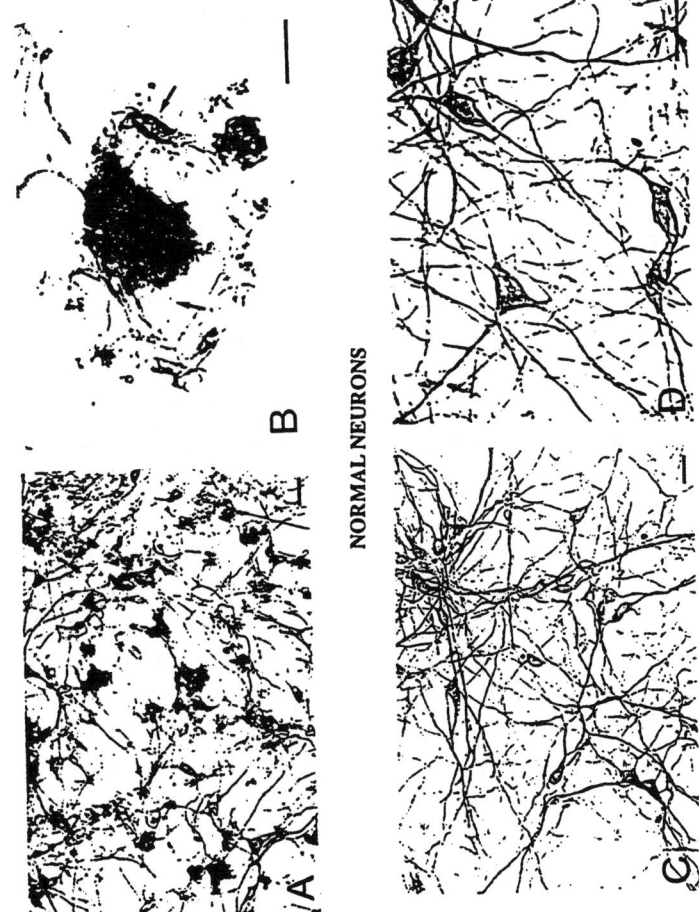

4). This is an actual photograph of a portion of the hippocampus. The cells which you see and which we are going to talk about are called pyramidal cells. Figure 5 shows how the size and complexity of these cells increase from the cell of a fetus to that of an adult human. In the fetus, the cell is small and contains an axon which connects it to the nerve terminal [branching at the top]. One single unbranched dendrite extends downward from the cell body. In contrast, the adult pyramidal neuron on the extreme right, contains a fully developed neuron with a large well defined cell body and many projections beneath it called dendrite trees. The axon proceeding away from the cell above the cell body is also branched and contains secondary projections. Later on we will discuss in more detail how these neurons communicate with each other. For now, I just want to point out that the axon of one neuron, which is referred to as the presynaptic neuron, releases chemicals that communicate with the dendritic end of a second neuron, referred to as the postsynaptic neuron. The space between the two neurons is called the synapse.

With this introduction we are now ready to answer the question, "What's wrong in an Alzheimer's brain?" In Figure 6, the top two panels [A and B] are photographs of tissue taken from a typical Alzheimer's brain. The bottom two panels [C and D] are photographs of tissue taken from normal human brain and correspond to the panels above them. When you read a Scientific American article or a New York Times article and it refers to senile plaque or amyloid plaque in Alzheimer's disease, what the article is referring to are those darkened areas in panels A and B which look like ink splotches. In color, these appear red because in order to visualize them they are stained with a dye called Congo Red. Panel B is a higher magnification of one of these plaques. The plaques are a clot of amyloid and tangled dead neurons. In panels C and D normal neurons are shown at the same magnification. Notice how the pyramidal cell bodies are visible as well as the many projections going off in all directions, allowing for communication among the neurons. In panel D the shape of the cells bodies can be seen more clearly.

We have just seen the pathology which leads to a diagnosis of Alzheimer's disease. The pathology was first described in 1907 by Alois Alzheimer who had been treating a severely demented patient and who upon postmortem examination of her brain, observed amyloid plaques and neurofibrillary tangles. The disease was named after Alzheimer and the plaques and tangles he observed became the hallmarks of the disease. Until recently the diagnosis of Alzheimer's Disease could only be made post mortem and was based on the following criteria: severe dementia, reduced brain mass, senile plaques containing an amyloid core, and neurofibrillary tangles. In 1990 at the Second International Conference on Alzheimer's Disease, there was no standardization of the dementia rating tests that were being used internationally to evaluate dementia patients, making it difficult for researchers to interpret each other's results. In addition, terms used to describe pathology were not standardized. For example, the protein core found in amyloid plaque was called beta-amyloid by some laboratories and A-beta by others. Soluble forms of beta-amyloid were referred to as s-amyloid and so on. This lack of standardized nomenclature still exists and you need to be aware of that when reading scientific articles. Currently, new diagnostic methods are being developed which allow for the diagnosis of Alzheimer's disease much earlier. Some of these methods involve brain scans using CT (computerized tomography) and SPECT (single photon emission CT), which are methods of visualizing the brain. Another method that is more familiar to you is MRI (magnetic resonance imaging) which is also a means of looking inside the body in a noninvasive way. By correlating brain size with degree of dementia and by tracking the rate of brain mass reduction and the increase of dementia over time (6 months or more) a diagnosis of probable Alzheimer's disease can be made. Alzheimer's disease is not simply forgetting things or events but is a rapid deterioration of both mental and physical function which often includes personality changes as well. The time span from diagnosis of the disease to the time of death is usually 7-9 years. It is estimated that by the year 2025, there will be over 3 million Alzheimer's patients in the United States and between 20-50 million patients world wide.

Senile dementia affects about 5% of the population over 65 years of age and 25% of those over 80. Fifty-five percent of the demented elderly have Alzheimer's disease. The rest of those with senile dementia may have suffered a stroke, experienced some form of head trauma, have a brain tumor or cerebrovasular disease. Sometimes those diagnosed with Parkinson's disease can also have dementia. For patients with dementia caused by cerebrovascular disease, drugs such as Captopril^R, a vasodilator, may help to improve memory. However, Captopril^R would have no benefit for a patient with Alzheimer's disease because the dementia in Alzheimer's is not due to narrowing of the cerebral arteries. Age associated memory impairment (AAMI) is another type of dementia which is not Alzheimer's disease. It is also referred to as "benign forgetfulness" and is not caused by disease.

What are some risk factors? Who is likely to get Alzheimer's disease? Some forms of Alzheimer's are inherited. This type of Alzheimer's is called familial Alzheimer's disease because it is found in families and is linked to mistakes in certain chromosomes, especially chromosome 21. This is the same chromosome which is linked to Down's Syndrome. If a member of your family has been diagnosed with Alzheimer's disease this does not mean that you will definitely get the disease, but it does mean that you may be predisposed to it. Very recently, Alan Roses, a researcher at Duke University, reported on the discovery of a blood protein, apolipoprotein E (ApoE) which plays a role in the rate of onset of Alzheimer's disease. He found that of the four variant gene types of ApoE, those who inherit the E_4 type from both parents have the highest risk of an early diagnosis (55-65 years) of Alzheimer's (E_4/E_4). On the contrary, those who inherit the E_2 type from both parents (E_2/E_2) are the least likely to be diagnosed as having Alzheimer's or will not exhibit clinical symptoms until very late in life (over 90). The E_3/E_3 combination, which most individuals have, falls somewhere in between and other combinations are possible. Other risk factors for Alzheimer's are age, head trauma (getting knocked unconscious or a severe blow to the head) and lack of education, which I will discuss in more detail later.

Now I want to discuss strategies toward finding an antidementia agent. How are we going to find agents to treat Alzheimer's disease? I want to begin by defining some terms for you. A neurotransmitter is a brain chemical that is produced by a specific nerve cell (neuron). Acetylcholine is the neurotransmitter produced by the cholinergic nerve cells in the hippocampus. The word cholinergic comes from acetylcholine which is made in the brain from the chemical choline. A cholinergic agonist is a molecule which mimics acetylcholine. A cholinergic antagonist is a molecule which blocks the action of acetylcholine. How do cholineric agonists and antagonists work in the brain? Think of it like a lock and key. The receptor (the part of a cell that receives the acetylcholine message and transmits the signal) can be thought of as a lock. The chemical that fits the lock and opens the door (allows the signal to be transmitted) is a cholinergic agonist. The chemical that fits the lock only partly (does not open the door but blocks other chemicals from entering) is a cholingergic antagonist or cholinergic blocker. The term blocker is often used to describe antagonists. Ones you may be familiar with are histamine blockers for treatment of allergies or beta blockers for the treatment of heart disease. Finally, I want to explain what is meant by the "cholinergic hypothesis". In the 80's a hypothesis was proposed that said that dementia was related to decreases in acetylcholine levels in the brain and that if these levels could be increased, memory would be restored. The rationale behind this hypothesis came from the fact that in Alzheimer's disease, cholinergic neurons are dying and consequently levels of acetylcholine production are reduced. It was reasoned that reduction of the neurotransmitter would lead to a reduction of signal transduction resulting in memory loss.

Figure 7 is a diagram of a cholinergic synapse with a presynaptic neuron on the left and part of a postsynaptic neuron on the right. Synthesis of acetylcholine (AcCh) occurs in the presynaptic neuron. Hydrolysis of excess acetylcholine (AcCh) by action of the enzyme acetylcholinesterase (AcEs) also occurs presynaptically. One of the earliest methods tried to test the "cholinergic hypothesis" was called precursor loading. In this approach, patients are given large doses of choline. The rationale is that choline will be converted to acetylcholine by reaction with acetylcoenzyme A in the presence of the enzyme, cholineacetyltransferase (CAT) in cholinergic neurons. A shortcoming of this approach is that ingestion of large doses of choline does not increase brain levels of choline. In addition, as cholinergic neurons die in Alzheimer's disease, it is the presynaptic neurons that die. This is the portion that contains the enzyme for making acetylcholine. Consequently, as the disease progresses, less and less acetylcholine is being produced. In conclusion, in precursor loading experiments, no measurable improvement in memory has been observed. In light of our current understanding that is not surprising. Contrary to presynaptic mechanisms, activation of cholinergic receptors of the muscarinic family (m_1 subtype) occurs on postsynaptic cholinergic neurons. These neurons remain intact in Alzheimer's.

ALZHEIMER'S DISEASE: CHOLINERGIC APPROACHES

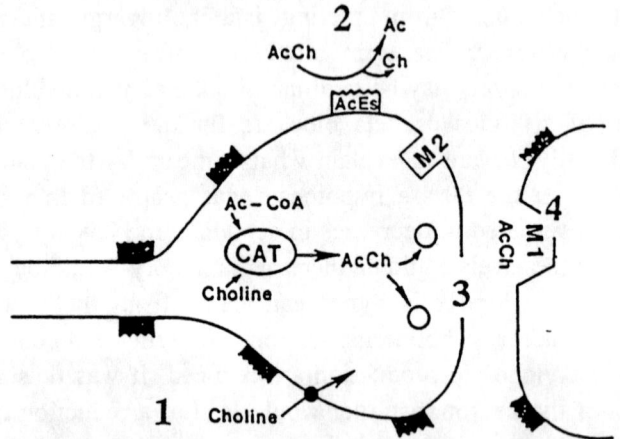

(1) PRECURSOR LOADING: GIVE CHOLINE OR LECITHIN

(2) CHOLINESTERASE INHIBITOR: i.e., COGNEX (marketed)

(3) ACETYLCHOLINE RELEASER: i.e., DUP-996 (in clinic)

(4) CHOLINERGIC AGONIST: ARECOLINE plus newer
       agents both in clinic and pre-clinical

Figure 7

The second approach is to give patients a cholinesterase inhibitor. This is the mechanism of the marketed drug Cognex^R (tacrine). Cholinesterase inhibitors work by inhibiting degradation of acetylcholine (AcCh) by the enzyme, acetylcholinesterase (AcEs). This degradation occurs in the synapse between cholinergic neurons. The rationale is that by inhibiting its degradation, existing acetylcholine will remain in the synapse and be available for binding to postsynaptic cholinergic receptors. Activation of these postsynaptic receptors (m_1), will restore signal transduction necessary for memory. The limitation of this approach is that it targets presynaptic cholinergic neurons which are dead or dying in Alzheimer's disease. In order for acetylcholinesterase inhibitors, such as Cognex^R, to be beneficial the drug must be given in mild to moderate cases of the disease where sufficient numbers of presynaptic neurons are still present. Cognex^R has been shown to have some efficacy in those patients which can tolerate the highest dose. Patients on Cognex^R must be closely monitored by a physician because liver enzyme elevation has been seen in some patients. In most cases, enzyme elevation is reversible when the drug is withdrawn. Newer acetylcholinesterase inhibitors are currently in the clinic. We can expect to see one or more of these marketed in the future, if efficacy equal to or better than Cognex^R can be demonstrated along with an improved safety profile. (Update September, 1996: Aricept^R (donepezil, E2020) found "approvable" by the FDA.)

A third approach is the use of an acetylcholine releaser drug, such as linopiridine (DUP-996) which is in clinical trials. Drugs of this type release stores of acetylcholine into the synapse. The acetylcholine stores are found on presynaptic neurons. As in the second approach, to be beneficial, acetylcholine releasers must be given in mild to moderate cases of the disease where sufficient numbers of presynaptic neurons are still functioning.

The fourth approach is a mechanism which targets the postsynaptic muscarinic receptors (m_1) which remain intact in Alzheimer's disease. Drugs of this type are called muscarinic (m_1) agonists. These drugs act as mimics of acetylcholine at postsynaptic muscarinic m_1 receptors. Drugs having this mechanism have been shown to improve memory loss due to Alzheimer's disease. In a study reported in 1992, researchers at the National Institute of Health evaluated the effect of a muscarinic agonist called arecoline on the memory of a small number of Alzheimer's patients. Arecoline is a natural product which is a non-selective muscarinic agonist similar in its activity to acetylcholine. In Alzheimer's patients given arecoline continuously intravenously (iv) over a two week period, improvement in cognitive measures was verified. Arecoline, itself, is not suitable for development into a drug due to its short half life. this is the reason that it was necessary to give the drug by the iv route of administration. Several major pharmaceutical companies presently have newer muscarinic agonists in clinical trials or in pre-clinical development. The most advanced of these is called xanomeline (Update October, 1996: Eli Lilly has completed Phase II clinical trials with xanomeline. The drug was found to show improvement in cognitive and some behavioral measures in mild to moderate Alzheimer's patients. The drug is rapidly metabolized when given orally. Phase II trails using a transdermal patch began in March, 1996. Other m_1 agonists in clinical trials include SB-202026 [SmithKline Beecham] and Talsaclidine [Boehringer Ingelheim]; Update March, 1998: SB-202026 is in Phase III clinical trials.)

The remainder of this lecture will be concerned with some of the strategies used by medicinal chemists to discover muscarine agonists (m_1) for the treatment of dementia. One place to begin is the structure of the endogenous neurotransmitter, acetylcholine. In the upper half of Figure 8 the molecular formula of acetylcholine is shown. Beneath it are two additional representations of acetylcholine. These representations are mirror images of each other. in them carbon-carbon bonds (C-C) and carbon-oxygen (C-O) bonds are represented by straight lines. When no atom is represented at the point where lines meet, the point is meant to represent carbon (C). Double lines represent bonds which are stronger than single bonds. A very important part of the structure of acetylcholine is the nitrogen atom attached to three methyl groups. [$+N(CH_3)_3$]. This nitrogen atom has a positive charge since the total number of bonds attached to nitrogen is four. This makes the nitrogen electron deficient. This group or a group electronically equivalent to it, is one of those necessary for binding to the active site of a muscarinic receptor. The term,

DRUG DESIGN

I . Endogenous Neurotransmitter

   A. Acetylcholine

$$\overset{+}{CH_3COOCH_2CH_2N(CH_3)_3}$$

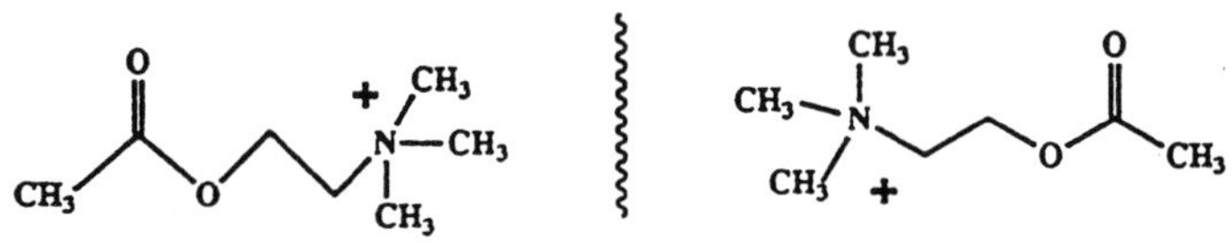

II. Natural Products as Muscarinic agonists

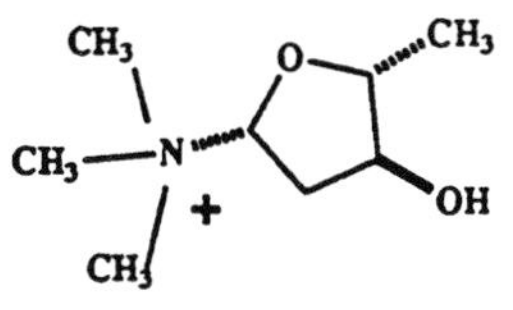

MUSCARINE

SOURCE: POISONOUS MUSHROOM

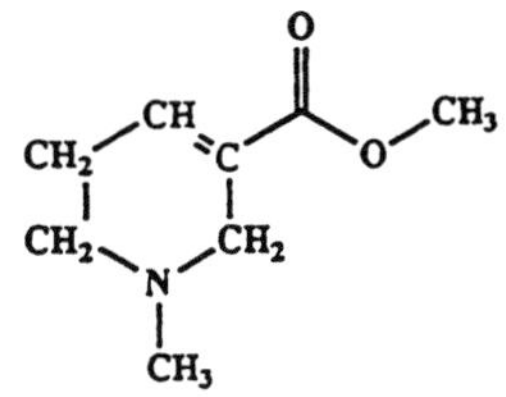

ARECOLINE

SOURCE: BETEL NUT

Figure 8

muscarinic receptor, is derived from the natural product, muscarine. It was the binding of this natural product to animal tissues which first identified these receptors. The source of muscarine is a poisonous mushroom. The other muscarine agonist derived from a natural product is arecoline. The source of arecoline is the betel nut. Both muscarine and arecoline are non-selective muscarinic agonists. Muscarine is a full agonist which means it fully actives muscarinic receptors. Arecoline is a partial agonist which means that it actives muscarinic receptors <100%. Since muscarinic receptors are found in both the brain and the periphery, a full non-selective agonist such as muscarine is expected to be toxic. Muscarinic receptors of the m_2 subtype are also found in cardiac tissue. A non-selective full m_2 agonist is expected to affect blood pressure and heart rate.

Because acetylcholine is a positively charged molecule, it is unable to cross the blood brain barrier from the periphery. As a result, acetylcholine itself, cannot be given as a drug. In the partial agonist, arecoline, the nitrogen atom is bonded to only three other carbon atoms resulting in an uncharged nitrogen atom. The neutral partial agonist, arecoline, can cross the blood brain barrier. However, because it contains an ester group [-(C=O)-OCH$_3$] it is easily degraded by acetylcholinesterase (AcEs) making it very short acting and not clinically useful. What is needed is a molecule which is more stable than an ester but with the same electronic properties. Such a molecule could substitute for the ester group at the active site of the receptor. Attachment of a trisubstituted amine to such a group should give an acetylcholine mimic (i.e., bioisostere, Figure 9).

Figure 10 shows some early drug candidates. These compounds, AF-102B (Abraham Fisher), SR 95639 (Sanofi Research) and AY-26514 (Wyeth-Ayerst Research), were reported to be muscarinic (m_1) receptor agonists. All three drugs bound to muscarinic m_1 receptors and were active in animal models of cognition at the time they were discovered (late 80's). In 1989 Dr. Thomas Bonner (National Institute of Health) reported that five subtypes of the human muscarinic receptor had been identified using cloning techniques. These receptor subtypes were designated m_1-m_5. It is now known that these subtypes are unevenly distributed throughout the brain and periphery (Figure 11). The m_1 receptor subtype is predominantly found in brain in the

DRUG DESIGN

BIOISOSTERES: molecules with different chemical structures but similar biological effects

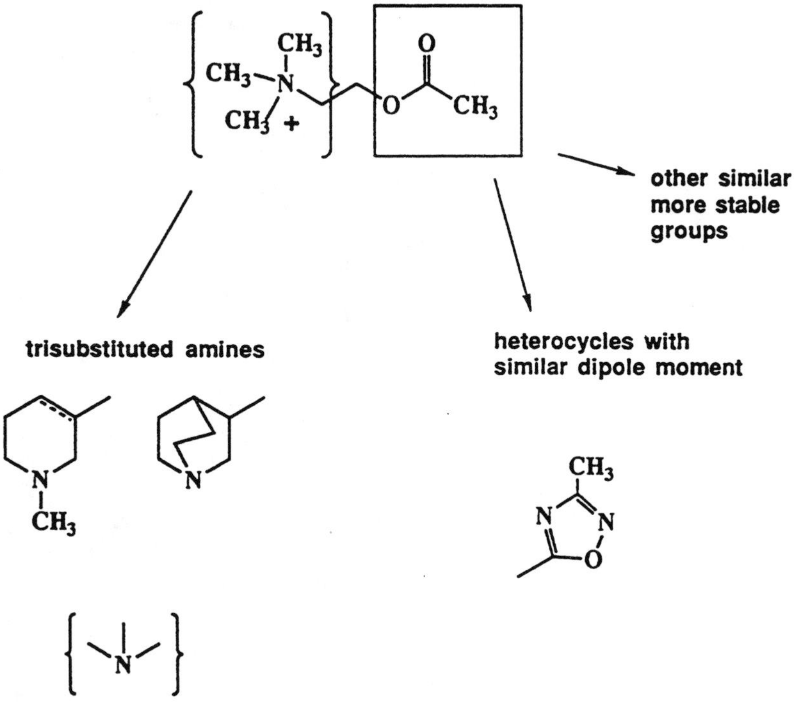

trisubstituted amines

other similar more stable groups

heterocycles with similar dipole moment

Figure 9

EARLY DRUG CANDIDATES

ACETYLCHOLINE MIMICS

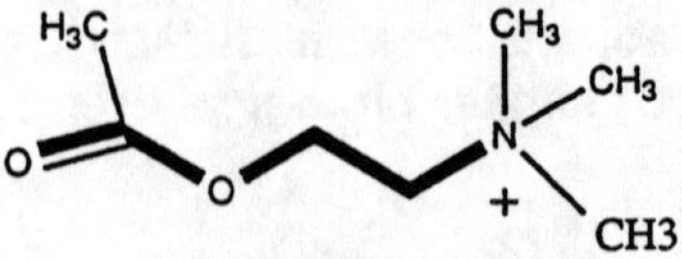

ACETYLCHOLINE

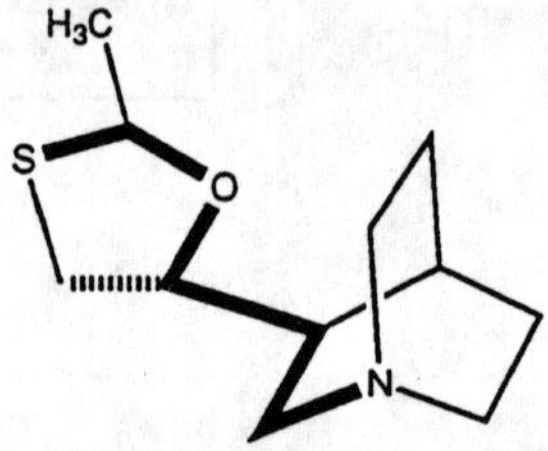

AF-102B

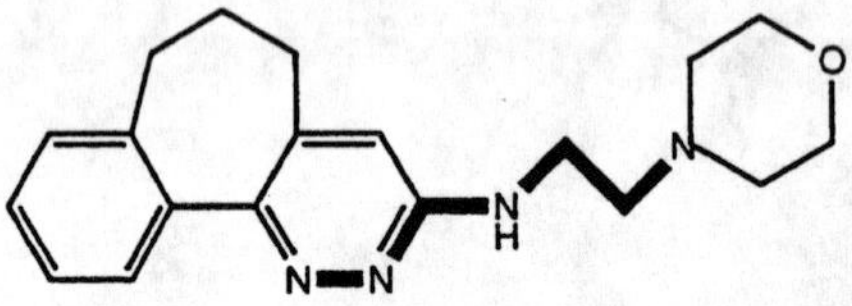

SR 95639

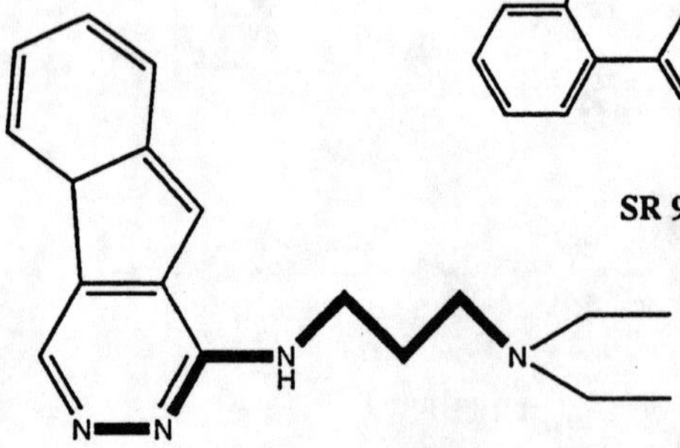

AY-26514

Figure 10

MOLECULAR BIOLOGY ADVANCE

1989-1990

•MOLECULAR STUDIES REVEALED MULTIPLE
(m1-m5) MUSCARINIC RECEPTORS

•THESE SUBTYPES ARE UNEVENLY DISTRIBUTED
THOUGHOUT THE BRAIN AND PERIPHERY

BRAIN

	M1	M2	M3	M4	M5
CORTEX	✓	✓	✓	✓	
HIPPOCAMPUS	✓	✓	✓	✓	
STRIATUM	✓		✓	✓	
CEREBELLUM		✓		✓	
BRAINSTEM		✓		✓	
SUBSTANTIA NIGRA					✓

PERIPHERY

	M1	M2	M3	M4	M5
HEART		✓		(✓)	
INTESTINES		✓	✓		
SALIVARY GLANDS	✓	✓	✓		
BLADDER		✓	✓		
PANCREAS			✓		
LACRIMAL GLANDS	✓		✓		

Figure 11

cortex. The m_2 subtype is found in the brainstem which controls bodily functions and in heart muscle in the periphery. The m_3 subtype is predominantly found in the salivary glands in the periphery. The function of the m_4 and m_5 receptors is still not well understood. The m_1 and m_2 receptor subtypes alone are capable of improving cognition. Acetylcholine or acetylcholine mimics can bind to postsynaptic m_1 receptors to activate signal transduction. m_2 Antagonists at presynaptic m_2 receptors can release endogenous acetylcholine from the presynaptic receptor. Using cloned human m_1 and m_2 receptors to study binding and agonism, it was found that AF-102B, SR 95639, and AY-26514 bound to both m_1 and m_2 receptors but were antagonists not m_1 agonists.

What is the muscarine receptor subtype and what does it look like? A muscarine receptor subtype is a protein which belongs to a family of proteins that thread their way across a cell membrane 7 times. Receptors of this type are coupled to other proteins within the cell called G-proteins. These G-proteins are part of the cell signaling mechanism. (Figure 12). The active binding site of muscarinic receptors is believed to be between transmembrane V and VI.

Molecular modeling is another name for computer-aided drug design. In order to design compounds which would not only bind to the m_1 receptor but activate it (m_1 agonists), the structural characteristics of known m_1 agonists were studied (Figure 13). The three compounds studied were carbachol (a standard full non-selective muscarinic agonist), arecoline, and a non-selective full muscarinic agonist from Merck (L-670, 207). The computer calculated regions in space near each molecule which would be most attracted to a proton (X, Y). It then rotated all flexible bonds to determine the most stable conformation of each molecule. Finally, it measured and plotted the distance between the nitrogen and the regions X and Y for each molecule. What was generated is referred to as a molecular electrostatic potential map (MEP). An MEP for the Merck compound is shown in Figure 14. By comparing the MEP's of various agonists and overlaying these, overlapping triangles are generated. Figure 15. This indicates that all the muscarinic agonists studied have three important structural characteristics occupying

MUSCARINIC RECEPTORS ARE PROTEINS WHICH
HAVE 7 TRANSMEMBRANE DOMAINS.

SPECIFIC AMINO ACIDS ARE NECESSARY FOR
BINDING OF SMALL MOLECULES TO THESE RECEPTORS

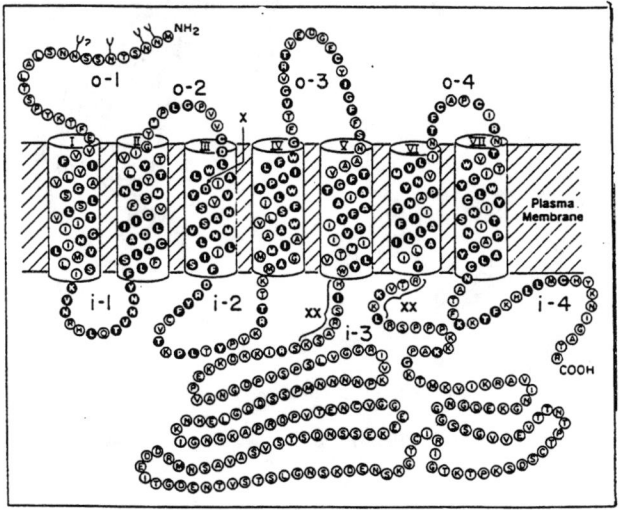

Figure 12 The Predicted Transmembrane Location and Amino Acid Sequence
of the M₂ Receptor.

MOLECULAR MODELING OF MUSCARINIC AGONISTS

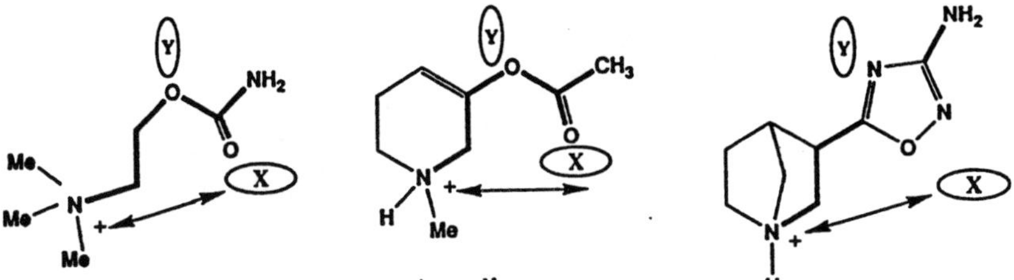

Carbachol

Arecoline

L-670,207

- Calculate regions in space most attracted to a proton (X, Y)

- Generate all possible conformations by rotating flexible bonds

- Measure and plot distance between quaternary nitrogen
  and center of calculated regions (X, Y)

Figure 13

MERCK OXADIAZOLE (L-670,207)

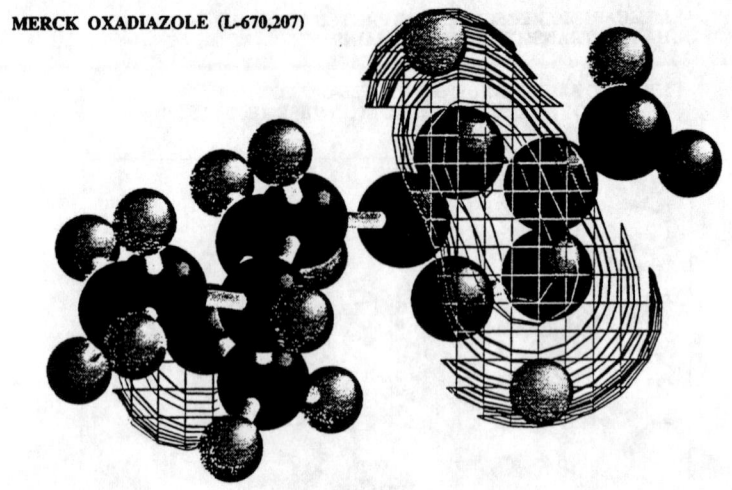

Molecular Electrostatic Potentials at Low Energy

Figure 14

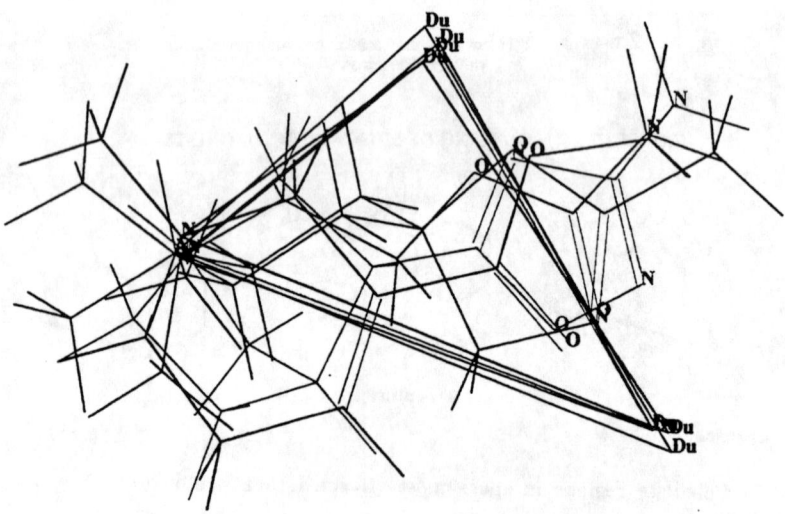

MEP fits for Carbachol(red), Merck(green), AF-150(magenta), and Arecoline(blue)

Figure 15

similar regions in space. Two of these points are hydrogen bonding sites [sites of high electron density where a proton (H ion) is attracted] and one of these is an ionic site (a nitrogen atom either possessing a positive charge or capable of acquiring one after binding to a proton.

An even more informative kind of computer-aided drug design is receptor modeling. Figure 16 is a computer-generated receptor model of the human m_1 receptor (left) and m_2 receptor (right). The seven transmembrane regions, shown in different colors, are helical. The presumed binding site in each receptor is shown in red sphere. The surprising structural similarity between the receptors suggests that differences in binding to these receptors should not be expected. So far only minor differences in binding affinities have been observed for all muscarinic agonists studied. Differences in functional activity between receptor subtypes are possible and have been found to be extremely sensitive to chemical structure. Figure 17 shows a model of the standard muscarinic agonist, carbachol bound in the m_1 receptor active site.

In summary, the kinds of drugs we have been discussing are designed to provide symptomatic relief for the memory loss which accompanies Alzheimer's disease. At the present time, only one marketed drug, Cognex^R, is available. This drug does provide some cognitive improvement in mild to moderate Alzheimer's cases but has the side effect of raising liver enzymes at high doses where the drug is most effective. In the near future, we expect to see marketed other acetylcholinesterase inhibitors with fewer side effects, acetylcholine releasing agents, and functionally selective m_1 agonists. (Update March, 1998: another acetylcholine esterase inhibitor, Arecept^R was marketed in 1997. This drug does not affect liver enzymes. Currently, slowing of cognitive decline is the treatment of choice. Female patients who have no prior history of cancer are often treated with estrogen, which has been shown to slow cognitive decline. Researchers are also investigating treatment with anti-oxidants, such as vitamin E, or treatment with antiinflammatory drugs which can cross the blood brain barrier, such as COX-2 inhibitors.)

Our next challenge is stopping the neurodegeneration of Alzheimer's. Some approaches to this are in understanding the role of

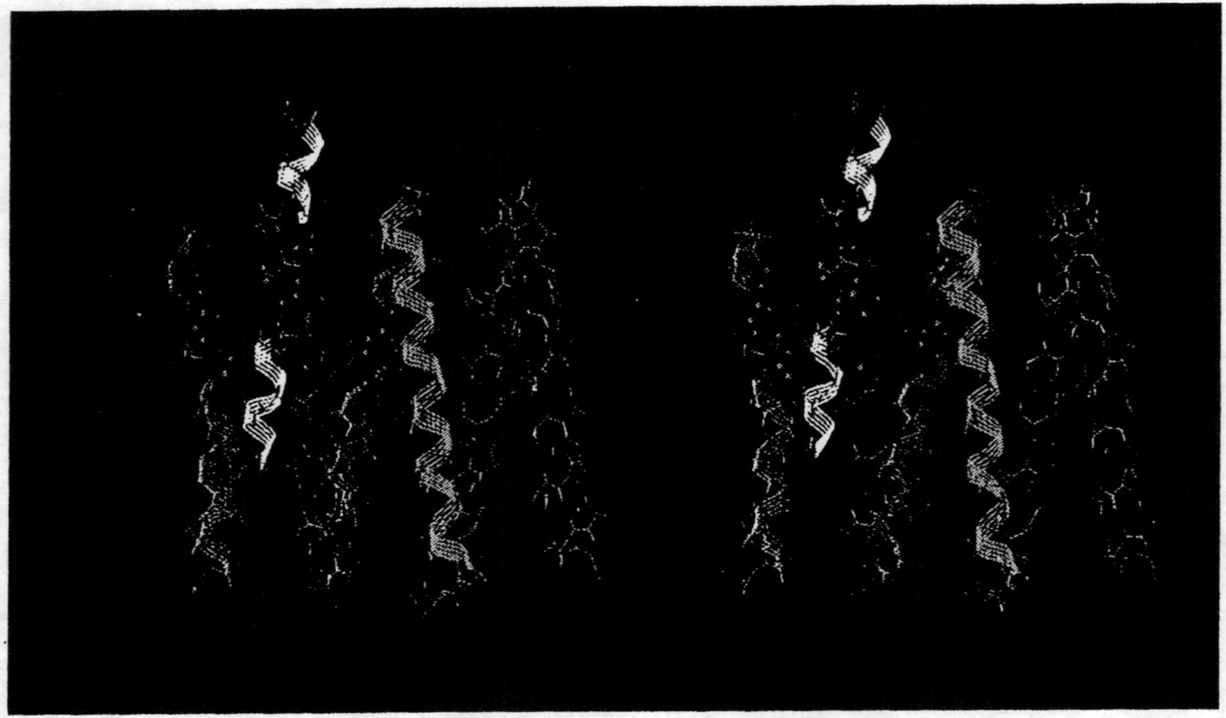

The seven trans–membrane helical regions of the m1 receptor (left) and the m2 receptor (right) with the presumed binding sites shown in solid red. (side–chains not shown)

Figure 16

Figure 17

beta-amyloid protein in Alzheimer's pathology. Are acetylcholine levels in brain related to beta-amyloid plaque formation? What other brain cells or molecules interact with beta-amyloid? What is the result? Another approach is understanding what causes cell death. What is the role of brain inflammation in Alzheimer's disease? Still another approach is a better understanding of the inherited basis of Alzheimer's. What gene defects are responsible for or predispose one to Alzheimer's? What is the role of apolipoprotein E (ApoE)? These and other related questions are under vigorous investigation in research laboratories around the world.

In closing I want to tell you about an article that appeared in the Journal of the American Medical Association on April 6, 1994. It was titled "Influence of Education and Occupation on the Incidence of Alzheimer's Disease." The conclusion of this study states:

> "The data suggest that increased educational and occupational attainment may reduce the risk of incident AD, either by decreasing ease of clinical detection of AD or by imparting a reserve that delays the onset of clinical manifestation."

What is meant by educational attainment is not the number of degrees earned but the continued exercising of the mind through the years.

ACKNOWLEDGMENTS

- The molecular modeling studies reported are the work of Dr. Reinhardt P. Stein, Structural Biology Department, Wyeth-Ayerst Research.
- Figures 1, 2, 3, and 4 are taken from "Structure of the Human Brain, A Photographic Atlas", S.R. deArmond, M.M. Fusco, M.M. Dewey, Third Edition, Oxford University Press, New York (1989).
- Figure 12 was taken from "Muscarinic Receptor Subtypes, Physiology and Clinical Implications" R.K. Goyal, *The New England Journal of Medicine*, vol. 321, p 1024 (Oct. 12, 1989).

- Figure 5 is taken from "Neurobiology," G.M. Shepherd, Second Edition, Oxford University Press, New York (1988), p. 615.

REFERENCES AND SUGGESTED READINGS

"Receptor Distribution in the Human and Animal Hippocampus: Focus on Muscarinic Acetylcholine Receptors," M.T. Vilaro, G. Mengod, G. Palacios, J.M. Palacios, *Hippocampus*, vol. 3, Special Issue Edition, Eds. R. Nitsch and T.G. Ohm, pp. 149-156 (1993).

"Tacrine for Alzheimer's Disease. Which patient? What dose?" Editorial by M.A. Winkler, *JAMA* vol. 271, no. 13, pp. 1023-1024 (April 6, 1993).

"A 30-week Randomized Controlled Trial of High-Dose Tacrine in Patients with Alzheimer's Disease," M.J. Knapp, D.S. Knopman, P.R. Solomon, W.W. Pendlebury, C.S. Davis, S.I. Gracon, *JAMA*, vol. 271, no. 13, pp. 985-991 (April 6, 1993).

"Acetylcholine release enhancing agents: potential therapeutics for Alzheimer's disease," R.J. Chorvat, R.A. Earl, R. Zaczek, *Drugs of the Future*, vol. 20, no. 11, pp. 1145-1162 (1995).

"Lipophilic muscarinic M_2 antagonists as potential drugs for cognitive disorders," H.N. Doods, *Drugs of the Future*, vol. 20, no. 2, pp. 157-164 (1995).

"Medicinal chemistry of muscarinic agonists: developments since 1990," E.K. Moltzen, B. Bjornholm, *Drugs of the Future*, vol. 20, no. 1, pp. 37-54 (1995).

"New Piece in Alzheimer's Puzzle," J. Travis, *Science*, vol. 261, pp. 828-829 (1993).

"Influence of Education and Occupation on the Incidence of Alzheimer's Disease," Y. Stern, B. Gurland, T.K. Tatemichi, M.X. Tang, D. Wilder, R. Mayeux, *JAMA*, vol. 271, no. 13, pp. 1004-1010 (April 6, 1994).

PIONEERS OF THE DIGITAL INFORMATION AGE: MORSE, SHANNON, KILBY, AND NOYCE

BRADLEY W. DICKINSON.

INTRODUCTION

This essay builds on the general theme of the relation of science to culture. In particular, I will focus on a topic that provides considerable insight into one particular role of engineering: contributing to the social and economic prosperity of our culture through the utilization of scientific knowledge. Both engineers and scientists employ a variety of mental processes to formulate models, and they follow the "scientific method" of inquiry to organize assemblies of empirical knowledge. Engineers, in their world, also rely on skills in *design*, the creative application of fundamental and empirical principles guided by performance criteria (possibly arising from conflicting goals) that include such crucial factors as economy, safety, reliability, manufacturability, and so on.

Of course many fields of engineering are very much involved in the relation of science to culture, and my particular fields of interest – signal, image, and video processing; telecommunications; and intelligent systems – certainly concern technologies that are at the heart of the "information age" of today and the foreseeable future. My topic may thus be viewed as a tribute to some of the giants on whose shoulders present-day researchers stand.

My viewpoint has been influenced substantially by my experience as a preceptor in Professor David Billington's course *Engineering in the Modern World*, where Princeton University students are engaged in reading, writing, and calculating in order to gain a broad appreciation of the major technological developments that have gone hand-in-hand with the political and social development of our country. The theme of *transformation* – technological, political and cultural – is emphasized in the course through the study of such major advances as the development of the steamboat, the telegraph, electric lighting, steel making, oil refining, the airplane, etc. Professor Billington's course examines their scientific bases, related societal issues, and their cultural influences. In this brief essay there will not be an opportunity to give more than a hint of some of the many related political and cultural facets. More can be found in the references as noted in the Further Reading section.

This essay has evolved from a lecture on the topic "The Scientist as a Man of His Times" that was presented as part of a program organized by Dr. David Peaslee. Thus, the focus on individuals of importance was adopted as a means of telling about some of the temporal developments of information technology. The names of the players are probably not equally familiar to a general reader: Samuel Morse, Claude Shannon, Jack Kilby and Robert Noyce. Such a short list omits many individuals, among them the intriguing case of characters who contributed to the early development of radio and television broadcasting – Marconi, Fessenden, DeForest, Armstrong, Farnsworth, Sarnoff. Thomas Edison and Alexander Graham Bell made familiar and important contributions as well. However, we have sharpened our focus to deal with *digital* information technology, in contrast to contributions that are fundamentally *analog* (audio telephony, audio recording, motion pictures, video recording, and the like).

Our four subjects are all pioneers of modern digital information technology, the technology of DBS (direct broadcast satellite) TV, DVD (digital video disk), HDTV (the soon-to-be-fielded system for high definition television broadcasting), CD (various compact disk systems for storage of audio, photographs, and multimedia), PCS (personal communications systems - the next generation of cellular systems), among others. For our purposes, we will think of digital information (discrete information) as that which consists of symbols (letters, digits, etc.), with sets of strings of symbols that convey meaning in an appropriate context (words, numbers, etc.).

Fundamental to digital information is its representation in primitive forms through *codes*, for example, Morse code, ASCII code, bar codes. These ultimately (in almost all applications) take the form (and may always do so in principle) of binary codes, strings of *bits*, a word coined by John Tukey from *bi*nary dig*its*. It should be recognized that the important issues of transforming analog information (acoustic signals, photographs, etc.) into digital formats have been addressed, and standards have been developed that have resulted in the digital "takeover" of virtually all forms of mass marketed high quality information.

Samuel Morse is well known for the invention of the telegraph. Claude Shannon developed algebraic techniques for analysis and design of digital information processes (as a Master's thesis at MIT!), and his best known work was the development of a mathematically rigorous and sophisticated approach to characterizing the performance limits of communication systems, from telegraphy to cellular telephony. Jack Kilby and Robert Noyce were the inventors of integrated circuits, a technology that has led directly to the ubiquity of fabulously powerful microcomputers and digital communication systems. It is notable how differences in these men reflect difference in the nature of developments in information technology over the last 160 years as well as differences in the American culture and society of which the individuals who provide technical leadership are an important part: Morse, a painter with a technical vision; Shannon, a mathematical genius who demonstrated fundamental principals of information processing; Kilby and Noyce, industrial researchers exploiting sophisticated physical principles and driven by technological demands.

SYSTEMS PIONEERING
DIGITAL INFORMATION REPRESENTATION AND
TRANSMISSION

Samuel F.B. Morse (1791-1872) was born in Charlestown, MA; he graduated from Yale in 1810. His ambitions as a painter were never fully realized and while teaching art at New York University he turned a hobby in electricity into a source of fame and fortune. The idea for using an electromagnet as the receiving device in a telegraph system came to Morse while returning by ship to America from Europe in 1832. Morse was a visionary who conceived of the entire telegraph system as a means of conveying information across vast distances. Lacking technical expertise, he was advised by a colleague, Professor Leonard G. Gale, who knew of Joseph Henry's work at Princeton on electromagnets, and aided by important technical consultations was led to a working demonstration system in 1835; Morse's receiver consisted of an electromagnet causing a pencil to mark on a moving strip of paper while the sending key was depressed, thus recording the motion of the key.

Commercial success of telegraphy took several years to achieve, and involved collaborations with Alfred Vail, Ezra Cornell, and Amos Kendall which were crucial in overcoming financial, political, and construction problems standing in the way of full realization of Morse's vision of the telegraph system – recognizable now as precursor of the information superhighway. Morse obtained a patent in 1840, and after years of lobbying Congress for funds, received federal money to set up a trial system between Baltimore and Washington in 1844. Despite its success, Congress was unwilling to establish a telegraph system as a public utility. By 1855, when the Western Union Telegraph Company was organized, the telecommunications industry was a major part of the private sector of the national economy.

Morse's most important invention was his code of dots and dashes, short and long duration key strokes that were used in combinations to represent the symbols of written information: letters and numerals (and spaces used to separate letters and words). Morse code is remarkably efficient in its use of short duration code strings to represent the most frequently occurring letters; furthermore, in conjunction with the redundancy inherent in written text, Morse code provides for substantial error correction capabilities that enable communication to be carried out even in the face of transmission and transcription errors. Indeed, telegraph operators learned to transcribe Morse code directly from the sounds of the clicking electromagnetic mechanism. (When wireless telegraphy was developed by Marconi, accurate transcription for weak signals, relative to background radio interference and noise, grew in significance.)

The significance of Morse code goes beyond its practical aspects. Conceptually, it provides a system of information representation and transmission by coded groups of binary, or on/off signals. Thus it was the first major embodiment of the kind of digital information systems that are pervasive in our present-day society.

One related aspect of Morse's telegraph system that exploits the digital representation of information is the electromagnetic relay. He recognized that the receiving electromagnet of one telegraph circuit could be used to operate the key of a second telegraph circuit located at a distance and powered by an independent source of electrical power (such as a battery). By chaining telegraph circuits together with relays, Morse devised the means to avoid any inherent limitations on the distance covered by the overall telegraph system. (A single telegraph circuit with a fixed power source is eventually limited on its ability to energize the receiving electromagnet because of energy dissipation arising from electrical resistance to current flowing in the wires, which grows proportionally to the wire length.) The relay is an example of a controlled switch, and while Morse would not have found it useful in the context of telegraph transmission systems, it functions as a type of primitive binary computing element. By interconnecting systems of relays in more elaborate ways, very general computing devices can be built – and were in the early decades of the twentieth century. With this in mind, we turn to our next pioneer.

ANALYTICAL PIONEERING
THE MATHEMATICAL FOUNDATIONS OF INFORMATION

The abstraction of mathematics has proved to be well-suited to many analysis and design problems in engineering. The major intellectual contributor to our understanding of information technology in its broadest sense has been Claude E. Shannon (1916–), who was born in Petoskey, MI and received undergraduate degrees in mathematics and in electrical engineering from the University of Michigan in 1936. He received the M.S. degree in Electrical Engineering from MIT in 1937 and the Ph.D. in Mathematics from MIT in 1940. After a post-doctoral appointment at the Institute for

Advanced Study in Princeton and a fruitful early career at Bell Laboratories, he returned to become an Electrical Engineering faculty member at MIT. Shannon's highly influential contributions may be generally characterized as discovering and exploring fundamental mathematical principles that govern communication and computation. Among his many honors is an honorary doctorate from Princeton University in 1962.

Shannon's contributions have been broad and profound. Indicative of his genius, his Master's thesis at MIT is widely regarded as the foundation of modern digital logic design. He introduced the mathematics of symbolic logic as the basis for analysis and design of digital computing circuits. When logic values TRUE and FALSE are represented by binary values 1 and 0, respectively, the combinational operations of OR and AND may be described by the corresponding binary addition and multiplication operations. The rules for rewriting and simplifying complex logical statements then reduce to the algebraic laws for binary expressions (e.g. the associative and distributive laws). Shannon's Ph.D. thesis was also remarkable, being a study of applications of symbolic algebra in genetics. (Technological advances in genetics were decades away, awaiting developments in the basic science of molecular biology.)

The field of electronic communications, stimulated by wartime research efforts (cryptography and radar), was poised for postwar expansion, and Shannon found it to be a fruitful one for his research efforts; he turned his attention to the problem of finding a mathematical formalism for information and communication. Using mathematical tools from the theory of probability, he authored a number of papers that defined the field now known as Information Theory.

Shannon's theory provided a conceptual framework for understanding communication systems such as Morse's telegraph: with no loss of generality, we may think of a communication system as comprised of a source coder (by which a source of information symbols is represented in the form of strings of bits), followed by a channel coder that transforms the bit strings into a coded form that renders the information resistant to corruption by the unpredictable effects of the transmission channel. Think of the transcription of written text into strings of Morse code symbols, which in turn are transmitted by current pulses on a telegraph wire or by electromagnetic energy pulses propagating through the atmosphere. What telegraph systems lacked from their inception was a means of providing error correction such as is found in modern modems used by personal computers and fax machines. (Consider the dire consequences if a Morse coded military message "You may not attack" is corrupted by a transmission error into the message "You may now attack".) Shannon's great contributions show that the conceptual framework (source coder followed by channel coder) leads directly to important quantitative characteristics of a communication system.

First, the capability for error correction is a fundamental, quantifiable property of a transmission channel; this is the implication of Shannon's remarkable Channel Coding Theorem. For sending messages across a communication channel involving noise (i.e. uncertain outcomes arising from unpredictable or random physical effects), there is an upper bound, the channel capacity (measured in bits per second), on the rate of error-free information transmission. At rates below channel capacity, perfect communication of information is possible through the use of error-correcting codes (in principle), while at rates above channel capacity there is an inevitable positive probability of transmission error for any communication system whatsoever. This result provides an absolute performance limit as a means of assessing efficiency of practical communication systems such as telegraph cables, wireless telegraphy, and even NASA's deep space probes. (Shannon also obtained results for analog communication channels such as radio broadcasting, telephones, etc.)

Second, the ultimate limit in the efficiency of representing a source of information symbols (such as alphabetic characters) by binary strings is determined by a quantity known as the information source entropy (measured in bits per information symbol). This is the substance of Shannon's Source Coding Theorem. Indeed, using fewer bits per symbol than the entropy will necessarily mean that exact recovery of the original source symbols will be impossible. (Shannon also elaborated on this theme by considering cases when corrupted recovery of the source is allowed; a fidelity measure is selected to penalize errors according to their significance to the ultimate user of the information. Such "lossy compression" is a key feature of the digital video communication systems such as DBS TV.)

Shannon's powerful results are timeless, in the sense that they capture ultimate limits on any conceivable technology. The technological developments that bridge the gap of time from Morse's telegraph to the present and beyond have been essential to the practical embodiment of digital information processing systems. Indeed, today's information age relies on advances in electronic technology of relatively recent vintage, and the last two pioneers to be discussed her were responsible for the key innovation: integrated circuits.

TECHNOLOGY PIONEERING ELECTRONICS TECHNOLOGY FOR THE INFORMATION AGE

The development of electronic technology suitable for the exploitation of information in digital form is a final aspect to be discussed. For nearly 100 years after Morse demonstrated his telegraph, there was little change in the technology (namely, electromechanical switches) used in digital information processing systems. In the 1940s, the first electronic digital computers were built using vacuum tube switching circuits; these suffered from high power consumption and relatively short time-between-failure characteristics. Only with the development of solid-state electronic devices, in particular the transistor invented at Bell Laboratories in 1948 by John Bardeen, Walter Brattain, and William Shockley (for which the 1956 Nobel Prize in Physics was awarded), did the commercial development of digital computers become a possibility, and then a reality.

The transistor must rank as one of the all-time most important inventions arising from the research sponsorship of big business. The integrated circuit (the IC "chip") became the ultimate realization of semiconductor electronics for digital information processing, and involved efforts undertaken by a large electronics manufacturer (Texas Instruments) in addition to more entrepreneurial efforts sponsored by venture capital (Fairchild Semiconductors, and later Intel Corporation).

One figure of merit that is used to provide a quantitative measure of digital processing technology improvements over time is the "power-delay product" for switching: take the power required to change the state of a controlled switch from OFF to ON and multiply it by the time it takes for the switch to change its state; better technology will reduce the power requirements and shorten the switching time. Semiconductor electronic switches outperform electromechanical switches and vacuum tube switches by significant factors. Increases in speed and reductions in power consumption are both important factors enabling higher performance digital processing systems to be built.

As digital system designs involve more and more switches, factors related to complexity of the digital systems need to be taken into consideration; for example, on average, how many other switches is a given switch likely to be connected to. In principle, there is no problem in designing highly interconnected systems, but there are practical limitations on performance and reliability. For example, to assemble a large digital system by wiring together switches according to a schematic diagram grows more difficult in proportion with the number of interconnections required, which is generally growing much faster than the number of switches. To be sure that every connection is correctly made, with good solder joints at each terminal, requires an accuracy of assembly that is achievable only at great time and expense. By the end of the 1950s, various efforts were under way to determine an effective way to carry out the assembly of complex digital systems. Integrated circuits provided the essential innovation that has been responsible for unleashing the potential of digital processing that has made ours the Digital Information Age. The start of this age was marked by a profound change in electronic technology where system-level considerations prevail, whereas earlier eras were dominated by device-level issues.

Interestingly, the origins of the integrated circuit are found in the work of two different men, Jack Kilby and Robert Noyce. Jack S. Kilby (1923–) was born in Jefferson City, MO. After receiving an undergraduate degree from Illinois in 1947 and an M.S. degree in Electrical Engineering from the University of Wisconsin in 1950 he joined Texas Instruments. Kilby's seminal idea, developed into a prototype circuit in late 1958, was to place all required circuit components (resistors and capacitors in addition to transistors) on a single piece of semiconductor material, suitably processed; this produced a so-called monolithic circuit. In Kilby's first demonstration circuits, the components were spaced relatively far apart, and interconnections between the components were comprised of thin gold wires carefully soldered in place with the aid of a microscope. Another man's ideas were soon to complete the innovation now known as the integrated circuit.

Robert N. Noyce 91927-1990) was born in Burlington, IA. His undergraduate education at Grinnell College was followed by a Ph.D. degree program in Physics at MIT. Noyce started his career in William Shockley's California transistor company, and by 1957 he and seven coworkers had so tired of Shockley's autocratic style of leadership that they quit and founded Fairchild Semiconductors; venture capital was provided by a young investment banker, Arthur Rock. (This pattern would repeat itself, with over fifty new semiconductor companies being started by Fairchild employees.) Noyce found himself in a management position, but maintained his technical activities as well. By 1959 he had conceptualized and then demonstrated the integrated circuit by using a patterned layer of aluminum wires laid down atop an insulated oxide on a silicon substrate. Even better, the processing steps used to add the wiring, vacuum deposition and photolithography, were the same ones required for fabricating devices on the substrate. Complete circuits could be built on a silicon chip!

In 1968 Noyce himself left Fairchild and co-founded Intel Corporation ("Integrated electronics") with Gordon Moore, who is best known for an empirical observation that has proved to be remarkably accurate as a characterization of the exponential growth of integrated circuit technology – Moore's Law: for state-of-the-art integrated circuits, the number of transistors that are fabricated on a chip doubles every eighteen months. Valid over the roughly three decade history of integrated circuits, some milestones of particular interest include two thousand transistors in Intel's first microprocessor, the 4004 (1971), one million transistors in the 80486 microprocessor (1990), and five million transistors in the Pentium Pro microprocessor (1996). The design of the 4004 microprocessor by Ted Hoff at Intel turned out to signal the start of the personal computer revolution. Today we can see that integrated circuit technology will continue to make the high technology of our Digital Information Age more accessible and more affordable.

FURTHER READING

The following bibliography provides a reading list that can be used by the reader interested in following up on the topics mentioned in this essay. They range from technological history to recent books on the roots and prospects for the Digital Information Age. The books by Lewis and by Cringely have corresponding public television productions: *Empire of the Air*, and *Triumph of the Nerds*, respectively; the latter traces the history of the personal computer revolution.

BIBLIOGRAPHY

David P. Billington (1996). "Henry, Morse, and The Telegraph." *The Innovators: The Engineering Pioneers Who Made America Modern*, Chapter 7, pp. 119-137. John Wiley, New York, NY.

John Bray (1995). *The Communications Miracle: The Telecommunications Pioneers from Morse to the Information Superhighway*. Plenum, New York, NY.

Ruth Schwartz Cowan (1996). *A Social History of American Technology*. Oxford University Press, New York, NY.

Robert X. Cringely (1996). *Accidental Empires*. Harper Collins, New York, NY.

George B. Dyson (1997). *Darwin Among the Machines*. Addison-Wesley, Reading, MA.

Richard Shelton Kirby, Sidney Withington, Arthur Burr Darling, and Frederich Gridley Kilgour (1956). "Electrical Engineering." *Engineering in History*, Chapter 11, pp. 336-351. McGraw-Hill, New York, NY.

Tom Lewis (1991). *Empire of the Air: The Men Who Made Radio*. Harper Collins, New York, NY.

Nicholas Negroponte (1995). *Being Digital*. New York: Alfred A. Knopf, New York, NY.

T.R. Reid (1984). *The Chip: How Two Americans Invented the Microchip and Launched a Revolution*. Simon and Schuster, New York, NY.

Claude Elwood Shannon: Collected Papers, edited by N.J.A. Sloane and Aaron D. Weyner (1993). The Institute of Electrical and Electronics Engineers, Inc., New York, NY.

Tom Wolfe (1983). "The Tinkerings of Robert Noyce. (How the Sun Rose on the Silicon Valley)." *Esquire*, December 1983, pp. 346-374.

A RELIGION FOR AGNOSTICS

THOMAS P. COOK

INTRODUCTION

The impulse to write this paper arose in recent years from my attendance at funeral services held in the Episcopal and Presbyterian Churches. Each institution still offers the traditional vision of God as a father who loves and cares for us and who, through the sacrifice of his only Son, Jesus, has saved all who believe in him for immortal life in a better world. These are comforting thoughts for those who can believe them. For reasons hereinafter given, I cannot subscribe to such orthodox Christianity, and therefore have sought a religion combining a scientific approach to the cosmos with all the sources of inspiration which can be found in the world around us.

Let me be the first to admit that my views may be replete with error, as the Catholic Church defines that term; that alongside the material world there may also be a spiritual one presided over by a personal God who rules both "Heaven and Earth". This God may also have become incarnate in the form of Jesus of Nazareth. Many devout friends of mine believe these doctrines, either on the authority of the Church or of the Bible, or because of a special revelation or experience. Who am I to say that *they* are in error? And what determines "truth" in the realm of religion?

The answer, as I see it, is that religious truth is relative to each individual. It depends for every person upon a number of factors including family background and tradition, religious education (or lack thereof), secular education, and perhaps most basically one's own psychology and emotional constitution. So for each of us, to borrow a phrase from Justice Holme's, the truth is simply "what I cannot help but believe."

For my own part, I cannot but believe in the natural world revealed to us by science. I am more impressed by objective knowledge of the universe and of what we experience around us than by subjective intuitions or revelations which may or may not correspond with reality.

This essay is addressed to those who, like myself, are unable to accept orthodox Christian dogma and who are seeking – or have found – other ways of faith which bring peace to the soul and gratitude for the gift of life. For whatever interest it may offer to my fellow heretics, I will attempt to define my concept of such terms as God, the soul, and immortality, and will expound a religious viewpoint which might be termed "The Faith of an Agnostic".

I. PROBLEMS WITH CHRISTIAN ORTHODOXY

Let me first elaborate on the main reasons why I cannot accept orthodox Christian dogma.

A. THERE IS NO OBJECTIVE EVIDENCE OF A SPIRITUAL WORLD SEPARATE FROM THE NATURAL ONE, AND THEREFORE I WOULD NOT COUNT ON ITS EXISTENCE.

Scientific studies of the brain and the other organs which in concert produce what one subjectively experiences as "the mind" have convinced me that the mind is simply the working of one's own organs – the electrical currents and chemical reactions – as experienced within. When the brain is anaesthetized, the mind goes blank. Neurologists have been able to locate in the brain the seat of many functions – vision, memory, creative thinking and the like. I cannot believe in the existence of a mind apart from a body, or a spiritual world separate from the natural. Body and mind, nature and spirit, all are parts of the same wonderful, mysterious flow of energy.

This "identity" viewpoint (as contrasted with "dualism") finds support in a profound book by the late physicist Heinz Pagels entitled *Dreams of Reason*. Concluding that "mental states and brain states are identical" and that "the mind and consciousness are a biological-material process explicable in terms of natural laws," Pagels reasons (p. 217):

> "If a mind is anything, it represents organized information and implies a memory exists... According to physics, any information transmission or signaling requires a change in energy... Any feature of our mental life that implies information is being processed must have material support."

To accept the unity of matter and spirit, it may be helpful to realize that matter is itself no more than energy organized in various combinations. Einstein's famous $E=MC^2$ has been empirically proven (unfortunately) by the atomic bomb. Thus, we must recognize that the world around us is not a fixed, static and ever-enduring structure, but rather a dynamic, ever-changing and evolutionary river of energy. When matter is broken down to its most basic particles, it becomes pure energy. Thus, we recognize that all "things" are temporary and transitory, and that the difference between matter and energy is one of form only.

From what we now know about the structure of the universe and the nature of life, it seems clear to me that each individual's subjective experiences take place within the framework of the physical body. Indeed, recent brain-scanning equipment has enabled the observer to pinpoint the area in the cerebral cortex where a particular thought was taking place. The evidence is now overwhelming that Dr. Pagels was correct in proclaiming that mental processes are physical.

What then shall we make of the content of those processes when they produce subjective revelations, conversions, imaginings and the like? We can conceive of a kingdom of heaven as described in the Book of Revelations and a hell as depicted in Dante's Inferno. Our capacity to imagine the attributes of a spiritual world seems to have no limit. When dealing with the realities of life and death, however, I cannot help but place more trust in scientific knowledge than in subjective intuition regarding a separate realm of the spirit.

B. A GOD OF LOVE DOES NOT GOVERN THE NATURAL WORLD.

Leaving aside the scientific approach to earthly existence, let us suppose that there is indeed another "realm of being" presided over by a God of love; it seems clear for another reason that such a god does not govern the physical universe. Consider the abundance of evil and cruelty in the world. A God that is both all-powerful in this cosmos and "loving" at the same time would not, in my view, tolerate so much suffering and pain to be inflicted on innocent and defenseless children and animals. Picture the brutality of a pack of wild dogs in Africa attacking an old zebra and tearing it to pieces bit by bit. Witness mankind, sometimes the cruelest animal of all, still torturing and killing each other over differences in politics or religion. Where was God's love in the Nazi holocaust?

C. THERE IS NO CONVINCING EVIDENCE THAT JESUS OF NAZARETH WAS THE "ONLY SON OF GOD".

The writers of gospels wrote of miracles performed by Jesus, the virgin birth, resurrection from the dead, and the ascension of Jesus into heaven. Similar stories appear in sacred literature of other religions, all of which would seem intended to make us into true believers. The marvels performed every day by natural forces, however, are far more compelling to me than tales of long ago which cannot be verified.

As for the authority of the Bible, it is an incomparable collection of books on numerous subjects including mythology, Jewish history, Jewish religious laws, the battles for their conquest of Palestine, books of wisdom such as Proverbs and Ecclesiastes, hymns and poems, sayings of the Prophets, four accounts of the life of Jesus written fifty years or more after his death, the letters of St. Paul and others, and finally the revelations of St. John. Much of this material, however, has little to do with helping us to understand and accept the world as it is and inspire us to make it better. The mythology of Genesis, the slaughter of the people whose land the Israelites wanted to seize, and the Jewish laws seem no more sacred than the Koran, the Hindu Vedas, the Meditations of Marcus Aurelius or the Prophet by Kahlil Gibran. All of these (and many others could be cited) are great and inspiring works. Furthermore, since the decision as to what books should be included in the Christian Bible was made by one group of early Christians while other Christians including the Gnostics and their gospels were left out, how can one honestly maintain that only the collection of books agreed upon by the prevailing sect was "the word of God"? While parts of the Bible inspire us to great heights, they were nevertheless written by human beings, and nothing human can be infallible.

The sufferings of this mortal life make us long for a better life elsewhere. But suppose there is no other world than this one in which we now live? One cannot disprove the existence of some creator or preserver separate from the universe itself; nor can one refute the possibility that such a supreme being may have one or more attributes of humanity, such as love, mercy or forgiveness. The agnostic simply takes the position that his religion should not rest merely on these possibilities, but rather upon what we can know and experience in relation to the natural world, of which human nature is a part. Instead of putting my faith in some hypothetical spiritual being or some unknowable entity separate from nature, I prefer to base my religion on the natural universe, and on what things in this life are of supreme value.

II. RELIGION DEFINED

Religion, as the term is used in this paper, deals with the relationship between the individual and the universe in which one finds one's self. It does not need to be organized: many people whom I would call deeply religious were not members of any church or similar organization. My great-grandfather John Bigelow was one of those who held fast to his "citizenship in heaven"; he intensively studied the Bible and its interpretation by the Swedish theologian

Swedenborg, and he devoted his life to improving the lot of his fellow man; but he was not a church-goer. His Faith was a private mater but with public consequences. (The reader may also recall Matthew, Chapter 6, in which Jesus advanced the idea of praying in the privacy of one's own room rather than in public as the hypocrites did.)

In my view, religion consists essentially of two parts:

1) the way a person sees himself in the scheme of things, and
2) the way to live, or what to do in our predicament. In short, it attempts to answer the basic questions: Where are we and what do we treasure?

This paper will deal first with the nature of the world around us, and will then offer some thoughts on how to live and to what purpose.

III. THE MYSTERY AND MARVEL OF THE COSMOS

I cannot but agree with Hamlet, when he said: "There are more things in heaven and earth, Horatio, than are dreamt of in your philosophy." What little we can and do know about the universe, however, reveals a supremely fascinating marvel whose mysteries have been gradually unfolding to science in recent times.

A. THE FORCES AND LAWS OF NATURE

I never cease to wonder at the fantastic organization of the cosmos, from sub-atomic particles held together by the so-called "strong force", to the hundred billion stars in our galaxy, which in turn is only one of 100 billion galaxies now observable, the most distant from us being some two billion light years away. The light from those distant bodies has taken 2,000,000,000 years to reach us! Who knows what lies beyond?

We have already noted another amazing fact – that all matter consists of energy in various combinations, from an atom of hydrogen to a human body. Even space itself may be energized. The energy asserts itself through such forces as electromagnetism and gravity, and it does so with such regularity that the operations of those forces are called "laws". Indeed, science has found in this universe an order and regularity which governs even such a complicated organization of elements as a human being. Each cell in our bodies contains the same set of genes as every other, enabling each cell to perform whatever task is demanded of it by the controls of the whole body: it can become a brain cell, a part of a bone, or a receptor cell in the retina of the eye. Looking at a scallop shell, one observes the beautiful fan shape in which the cells arranged themselves. At the atomic level the elements invariably reat with each other in accordance with the laws of chemistry and physics.

All science proceeds on the assumption that the natural order is not subject to interference by some "super-natural" force or entity; if some event or phenomenon cannot yet be explained by science, it is not viewed as a supernatural "miracle" but rather as a subject for further inquiry.

Albert Einstein has written convincingly on this subject, discarding belief in a personal god and recovering the natural cosmos instead. I quote here from an essay of his on science and religion:

> "The more a man is imbued with the ordered regularity of all events, the firmer becomes this conviction that there is no room left... for causes of a different nature."

It is this order in nature in which an agnostic can trust and can put his "faith". Of course, the forces of nature constantly cause vicissitudes, variations and changes on earth in the heavens: fair weather and foul, accidents and illness, mutations in genes, and the collapse of stars into black holes. Yet within this framework of contingencies, when it comes to solving life's problems, we can rely on the order in nature much better than on a possible super-natural being who will intervene on one's behalf if one prays for such intervention.

B. LIFE.

Whatever may be the ultimate force or forces or elements underlying all nature, the miracle is that from these basic forces and particles have evolved all other natural elements, and these in turn have combined to form life in all its diversity. The evidence now shows beyond doubt that the progression from the inorganic to the organic came through the natural evolution of more complex molecules from simpler ones, eventually producing a molecule which could reproduce itself.

According to Professor Edward Wilson of Harvard, whole life began about 3.8 billion years ago from prebiotic organic molecules, it was only in the past 500 million years that there evolved the millions of species that have existed to date. During the same time the turnover of species was nearly total; more than 99 percent of all species that ever lived in each period perished, and thousands more are becoming extinct every year as new ones emerge.

Thus, life forms one stage in the endless cycle of combination, dissolution and recombination of natural elements, which in turn consist of energy in diverse organizations. Rather than "reducing" life to the Biblical "Dust", let us venerate this "dust" and all natural forces as the generators of life. The soil in the garden, the manure that fertilizes it, the sun and rain which make life possible – these are the miracle-workers which have evolved and sustain living things.

IV. THE INDIVIDUAL AS PART OF NATURE

A. LIFE AND DEATH

According to my recollection, Einstein once said: "I am one small part of nature." That is my viewpoint. Each of us is the product of the forces of heredity and environment – the genes which have been bequeathed to us by our ancestors and the surrounds, human and otherwise, which helped to form each individual character. The forces of nature – electrical, chemical, biological and social – constantly operate on us from both within and without. At death each of us dissolves back into the simpler elements which had combined to give us life and which now will be available to sustain new life. As Kahlil

Gibran put it in *The Prophet*, " Life and death are one, even as the river and the sea are One."

B. ALL THINGS INCLUDING PEOPLE ARE EVENTS.

In view of the transitory character of life, I like to think of myself as an event or complex of events rather than an object. Everything changes, some changes being much more rapid than others; and eventually even the whole solar system will pass away. Every organism goes through the process of growth, maturity, decay and return to the inorganic reservoir of all life. Each of us may be likened to a sand castle on the beach; it flourishes until the next high tide comes and dissolves it back into the sand from which it was made. The same sand then awaits the next hand to form it into another sand castle. Or one might liken one's self to a drop of rain; it arose from the ocean as vapor, eventually condensed into a drop, then fell back into the ocean – its origin and destination. So every person, like a sand castle or a drop of rain, exists for awhile as an individual, but eventually returns to the source from which it came and of which it will always be part.

C. BODY AND SOUL – ONE AND THE SAME.

The soul, as I view it, consists simply of the entire thoughts, feelings and actions of a particular life. For the reasons stated earlier (pp. 3-4), all these elements of the soul have a physical basis in the body. The human body, in turn, is not a fixed entity, but a continually changing, growing and evolving aggregation of chemical elements which nature has miraculously developed from simple molecules into an organism of unbelievable complexity. Thus, soul and body are simply different aspects of the same wonderful process called life. When the body dissolves at death, the soul ceases its strivings and rests in peace.

V. GOD AND IMMORTALITY.

At this point let us consider two other concepts which occupy a central place in orthodox Christianity and can play a useful part in agnostic religion: the concept of God and the idea of immortality.

A. "GOD" DEFINED.

Although many have rejected the use of the term "God" because it has meant so many different things to different people, I believe the word is useful in any essay on religion; therefore, I want to define the sense in which it appears in this paper.

Webster's Dictionary defines God as "the Supreme Being, the eternal and infinite Spirit, Creator and Soverign of the Universe." This concept implies something above and apart from nature, but responsible for its existence – an ultimate, spiritual, power, separate from the physical universe which it created. This dualistic philosophy likewise views man's spirit or soul as separate from his physical body; the latter dies, but the soul somehow continues to live on in the spiritual world. As Longfellow's "Psalm of Life" put it, "Dust thou art, to dust returneth, was not spoken of the soul."

This dualistic view is unacceptable to me for the reasons given earlier in this paper (pages 3-5). Instead, I conceive of God not as a spirit apart from this universe, but rather as the creative, evolutionary power which pervades – indeed is – our cosmos, and perhaps more.

The word "God" in my vocabulary means the ultimate ground of all being – the source of everything, past, present and future – the "one tremendous whole", as Alexander Pope described it. It is the natural force or combination of forces which moves all things. I use the word "God" because it is not only a short and simple term for the most profound concept of which man is capable; it also connotes the awe, wonder and reference which most of us experience when contemplating the ultimate mystery of all existence.

As a matter of pure speculation, according to the great English physicist Stephen Hawking, there may be other universes besides ours, and so there may be an ultimate power which created or evolved all of them including ours; such an ultimate power could be called

God. If that is the case, however, the creative forces of our own cosmos may still be a part or a manifestation of the ultimate power which is in and through all.

In my view, there is no question whether God exists; the universe exists, and either it is God and is self-generating, or it must have been brought into existence by God as the "prime mover" which created everything including the cosmos and its laws. It makes little practical difference whether God is the universe or whether He created it because, as Hawking and Einstein have observed, the universe continues to evolve in accordance with its (or God's) laws and (as far as we can tell) without interference by any outside or "supernatural" force.

The generating, evolutionary power of our universe can be observed everywhere, and it alone suffices to evoke our most profound religious feeling. So, whether the universe is the ultimate creator or whether its forces are a manifestation or product of a further creative power, my preference is to use the word God as synonymous with the universe in which we live, while leaving open the possibility that our universe may be part of or controlled by an even greater but unknowable power.

B. WHAT IS IMMORTAL.

Nature has bred in each of us the urge to survive or the will to live, and a brain in which we experience consciousness (except when we are asleep or anesthetized). This combination has led many to believe that somehow each of us will go on "living" after the body has died. There is also a tendency to project our own thoughts, values and humanness onto the rest of nature (the Greek cosmology for example). This tendency to anthropomorphize leads many to project human qualities onto God – the ultimate reality. They conceive of this God as a person and they create images of an after-life such as in Dante's Paradiso or in many Christian hymns. All these human ways, however, must be viewed in the light of our objective and scientific knowledge of the universe as a whole, and of how life has developed as part of this universe. This forces us to realize that individual human beings – marvelous as we are – are the creatures of the forces of nature; that we should not expect to live on after death except as we have been parts of the everlasting processes of natural evolution. The individual is but a temporary manifestation or creature of the all-embracing power that I call God. Our ideals and spiritual lives are indeed real, but they belong within the context of a natural universe that alone endures.

It follows from the foregoing that every life is as immortal as the elements which compose it and which, in one form or another, always will be a part of nature. We are immortal to the extent that the universe is or may be immortal. Each life is an event or series of events in the progress of cosmic forces and particles which will continue to the end of time, or to infinity if time never ends.

The Prophet Isaiah had it right when he said:

"All flesh is grass, and all the goodness thereof is as the flower of the field. The grass withereth, the flower fadeth; but the word of our God shall stand forever."

VI. HOW TO LIVE AND WHY

If my views of the universe and of human existence are accepted, then we face such questions as: Why is life worth while, and how should we try to live it?

To all such questions there is a simple answer as stated by Justice Holmes: "Life is an end in itself, and the only question as to whether it is worth living is whether you have enough of it." To live is to function, he said, to use one's powers; and one of the joys of life is "to put out one's power in some natural or useful or harmless way." And there are so many other joys to be had, from good bread and drink to love and marriage (see Ecclesiastes, Chapter 9); doing a good job in whatever calling one pursues; creating and enjoying beauty in all its forms; firm and mutual friendship; producing and raising children; contributing to the betterment of the world around us; and finally, the "cosmic religious feeling" which one can derive from contemplating the grandeur and ultimate mystery of the universe and endeavoring to know as much as one can about it. Einstein described this feeling as "a rapturous amazement at the harmony of natural law, which reveals an intelligence of such superiority that, compared with it, all the systematic thinking and acting of human beings is an utterly insignificant reflection."

In a sermon delivered in the Princeton University Chapel in 1966, George Kennan remarked on how much in life is good:

> "There is, after all, this marvelous earth around us, for anyone who wants to see it and to sense it: There is the sky, the clouds, the sunshine, the wilderness; there are growing things and wildlife; there is the sea; And there is the mysterious enrichment that is to be had from being near all these things, from being near to Nature and alone with her, from feeling one's self a small part of her, from participating – whether as a gardener or a camper or a sailor or what you will – in the rhythm or her vast processes as they have existed and still exist wherever man's interference has not corrupted them."

Of course we all suffer pains and frustrations – some to a far greater degree than others. Mortal existence can become such a burden, as where a terminally ill cancer patient is enduring hopeless pain, that I see no valid reason not to end that life if the sufferer so desires, and why he or she should not receive assistance to that end if requested. I find no merit in the argument that because God gave us life, we should not terminate it when the pain has become intolerable and the prognosis hopeless.

Further as to how to live in the human predicament, let me record two brief passages which may be helpful. The first is Lord David Cecil's interpretation of Joseph Conrad's view of life:

> "What one lives for may be uncertain, but how one lives is not... Man should life nobly though he does not see any practical reason for it, simply because in the mysterious inexplicable mixture of beauty and ugliness, virtue and baseness, in which he finds himself, he must want to be on the side of the beautiful and the virtuous."

The other is another quote from Albert Einstein:

> "A human being is a part of the whole, called by
> us Universe", a part limited in time and space. He
> experiences himself, his thoughts and feelings as
> something separated from the rest – a kid of optical
> delusion of his consciousness. This delusion is a kind
> of prison for us, restricting us to our personal desires
> and to affection for a few persons nearest to us. Our
> task must be to free ourselves from this prison by
> widening our circle of compassion to embrace all
> living creatures and the whole nature in its beauty.
> Nobody is able to achieve this completely, but the
> striving for such achievement is in itself a part of the
> liberation and a foundation for inner security."

What about those who, because or horrible abuse in childhood or
from other causes, have been so warped in their development that
they have lost their capacity to love and their appreciation of "the
beautiful and the virtuous", and have been driven to insanity, violent
crime, or both? As long as a life remains in this condition, it would
seem to have little value. Sometimes, however, a miracle happens and
through therapy or other help a lost soul is saved. The more fortunate
of us should support efforts to salvage the wrecks of society when
possible, and more importantly, to forestall or improve the conditions
that lead to perdition.

VII. THE SUPREME VALUES: FAITH, HOPE AND LOVE

In I Corinthians, Ch. 13, St. Paul delivered his famous exhortation
on the three supreme virtues of faith, hope and love. Let me discourse
briefly on those subjects.

My faith, if it can be called such, is – very simply – in God, the
ground of all existence. I feel complete confidence in and
identification with the power in which "we live and move and have
our being". The creative process works in and through each of us. As
the Psalm says: "It is He that has made us, and not we ourselves."

What each of us thinks and does is the result of the force or forces which continually shape our lives. Each life cannot but play the role assigned to it and must accept whatever fate befalls it.

This deterministic philosophy raises the time-honored question whether human beings have "free will" and can therefore be held responsible for their actions. In *Dreams of Reason*, Pagels maintains (pp. 219-221) that while in principle the mind and consciousness are biological-material processes explicable in terms of the natural laws of electricity and chemistry, it is for practical reasons that we have to treat people as free and responsible moral agents. At the same time we hope through education, legal sanctions and other influences to condition people to be law-abiding and civilized. Such conditioning proceeds on the assumption that while an individual may feel free in making choices, what he or she chooses will be determined by all that has gone into forming that person's character. In the light of that fact, we should feel sympathy for those in trouble. "There but for the grace of God go I."

Providence, as our forefathers called it, favors some and is hard on others. Yet those who suffer more than most can have one consolation; as Marcus Aurelius put it in his Meditation No. IV:

> "Remember too on every occasion which leads
> thee to vexation to apply this principle: not that this is
> a misfortune, but that to bear it nobly is good
> fortune."

How does this faith of the agnostic or the stoic relate to hope? Mankind has lived through countless periods of intense suffering and utter destruction – the black plagues of medieval times, mass starvation, brutal wars in which entire populations of vanquished have been wiped out, and today the human population explosion with its consequent ecological disasters. In history's darkest hours many must have abandoned all hope for the future.

Yet the human race has survived and many dedicated people are working hard for a better world. We are striving to level off population growth, to preserve other forms of life from extinction, to protect our environment from further desecration, and to promote peace and human rights around the globe. The regenerative power of nature continues to reassert itself, as the phoenix rises from the ashes. Fifty years ago, when Hitler's military machine and Stalin's repressive dictatorship were "riding high", who could have foreseen the formation of the European economic community and reforms in the Soviet Union under Gorbachev? Nor should we overlook the tremendous advances in medical science, molecular biology, computers and space exploration, all of which hold such potential for a better and richer life for us and our descendants. The course of history provides indeed a firm foundation for hope.

Kennan's 1966 Sermon above referred to was entitled "Why Do I Hope?" He reasoned that in the grim present (the nuclear arms race was then on at full speed) no one could predict the future course of events; that in any case we must "fight the good fight" while we can; and that, most importantly, "there is in this human existence the phenomenon of love."

This brings me to the last – and greatest – of St. Paul's trilogy. We find the spirit of love, sympathy and concern built into human nature in spite of all its failings. We are born to love and to care, much as selfishness, greed, hate and fear may interfere. And love that has emanated from the lives of others, and the examples of their dedication and sacrifices for the benefit of their fellows, have a powerful effect on us. We may not believe that Jesus was God; but the story of his ultimate sacrifice for the religion of love he believed in may well be the most moving story every told.

For many of us agnostics, Jesus of Nazareth has by his teachings and example become the symbol of love and dedication – the qualities which we most revere and look to for guiding our own lives. In my view, true Christianity is very simply a religion of love. Jesus himself said as much when he condensed his religion into two basic precepts: Thou shalt loves the Lord thy God with all thy heart and soul, and Thou shalt love thy neighbor as thyself.

Agnostics can hitch their wagons to this same spirit of love – the "constructive principle" – which they observe and experience in nature, including human nature. It is that spirit which continually strives to make the world better, to reduce pain and oppression, and to produce as much goodness and beauty as possible.

I once heard a Unitarian Minister deliver a sermon in which he summed up his religion as closely akin to basic Christianity as I have just defined it; he stated his two precepts in succinct but profound words on which I cannot improve: think truth, and live love.

THE PLACE OF SCIENCE IN THE EDUCATION OF NON-SCIENTISTS[1]

F.H. WESTHEIMER

INTRODUCTION

Many major universities, including those in the "Ivy League", demand very little instruction in science of their undergraduates. Although technology and its underlying science dominate modern society, most students do not voluntarily choose instruction in science[2] , and most faculties do not require it of them. Why? Or more precisely why not? Are science courses too difficult? Too dull? Does anyone seriously doubt their importance?

My thesis, my hypothesis is that the problem arises in large part because of the vertical nature of science where each subject depends critically on prior learning, even at the college level. Vertical learning is, of course, the pattern in primary school; for example, reading at the third grade level depends critically on reading skills from the second grade. But this sequence of dependency does not continue so far in the humanities; reading at something like the tenth grade level suffices to allow students to study history and literature, whereas the vertical nature of learning in science continues well into college. The importance of the vertical nature of learning in the sciences is the subject of this article.

---

[1] Earlier versions of this paper were presented in December, 1993 at a conference sponsored by *Daedalus* and a meeting in June, 1994 of a group from Princeton's class of 1943.

[2] Anecdotal evidence: The Princeton undergraduate who operated the slide projector at the June conference explained that he was good at science, but if Princeton had demanded three half-courses in science rather than two for graduation, he would have gone to Harvard instead.

TECHNOLOGY

The technology that characterizes our society is obvious. We take for granted electric motors and electric lights, automobiles and airplanes, telephones, videos, computers, a wide variety of metals, plastics and synthetic fibers, anesthetics, antibiotics, hormones and vitamins, grapefruit and nectarines, and much else that arises from technology. "Nearly every article we touch is colored, coated, protected, stabilized or otherwise modified by synthetic chemicals. Almost all detergents, fertilizers, lubricants, fuels (including 'ordinary' gasoline), antifreezes, disinfectants, pesticides, cosmetics, adhesives, solid-state devices, energy converters (fuel-cells, magnets, lasers), and thousands of other products are constituted wholly or in part of synthetics."[3] In the course of history, much technology preceded science, but in most cases today science has caught up, and now leads technology.

In our society, business men and women, and lawyers head most large corporations, and to a large extent "run" our country. Non-technical managers command the engineers and scientists who make the wheels turn. Don't the executives want to know something about the nuts and bolts in our economy? Don't they need to?

SCIENCE

Many of the proudest intellectual generalizations of mankind arise from consideration of the findings of science, and might reasonably command a place in education. Examples: the Copernican solar system, gravitation, the ideas of convergent and divergent series, the chemical elements and the periodic table of those elements, the germ theory of disease, evolution, the laws of genetics and the biochemical basis of these laws, and much else. Modern science can explain what makes the sun shine and a large part of what makes plants grow. At a less fundamental level, modern biochemistry has explained the mechanism by which penicillin combats bacterial

---

[3] *Chemistry: Opportunities and Needs*, National Academy of Sciences Publication, no. 1292 (1965).

disease. Medical science has discovered viruses, and molecular biologists have illuminated the structure of viruses, and explained why they are immune to penicillin and other antibiotics. Chemists have started to work out prebiotic chemistry, that is, the chemistry that produced the molecules of life before there was life on earth – the molecules from which life evolved.

No one should minimize the pleasure in learning something of history, English and other literature, economics, social studies, art and music. But although the concepts of justice, law, religion, beauty in art and music, and the historical record of mankind in many civilizations are proud intellectual achievements, they certainly do not overwhelm those of science. Does not an education demand a reasonable amount of both?

One might think that students would insist on learning about the practical and the theoretical advances in science and technology, but many of them don't. One might think that university administrators and faculty would insist that all students have a modest acquaintance with much of the material and many of the concepts mentioned above. They don't. The universities provide numerous courses that teach most of the ideas and some of the practical maters of science. But these courses are taken almost exclusively by those who intend to make careers in science, engineering, and medicine. Fortunately, something in the solar system and the germ theory of disease is usually taught in high school, and the controversy over evolution has brought that concept to the attention of most educated people, although only a few are acquainted with the spectacular way in which the theory has been reinforced by the findings of biochemistry – perhaps because so few people know much biochemistry. In later sections, some tentative suggestions are offered to explain why the prospect of correcting the deficit in science is so bleak.

CORE CURRICULA

Core curricula have been instituted in many colleges and universities, and one of the purposes of such curricula is to supply non-scientists with a background in science. Despite an occasional bright spot, and much sincere effort, most of these curricula are grossly inadequate.[4] For example, the Core Curriculum at Harvard requires each student to register for two special half-courses in science – approximately 6% of his or her total curriculum. Is it possible to maintain, seriously, that the items listed above constitute only 6% of a general education? And Harvard is not alone in this strange deformation of the curriculum; Princeton is only slightly better; Yale and Columbia used to have similar curricula, but both have recently marginally increased their requirements for graduation to three half courses in science and math. Except for universities specifically devoted to science and technology, such as MIT and the California Institute of Technology, such curricula are common.

In 1849-1850, every student at Harvard was required to take a course in mathematics or science in every semester of every year. Presumably other American colleges had similar programs. Since 1850, science has expanded enormously; the vast majority of the science we know has been discovered since then. As typical examples, the transistor, the laser and the digital computer were invented, nitrocellulose has replaced gunpowder, the structural theory of organic chemistry and the *Origin of Species* were put forward, quantum mechanics and relativity were introduced, the double helix was found, nuclear power plants and nuclear bombs were invented; and so on and so on. How is it possible that any university should so greatly diminish the science taught to its graduates during a time when the practical and intellectual importance of science have exploded?

Dozens of colleges and universities have debated the question of curricula, dozens of curricula have been tried with more or less success, but essentially all are alike in that they de-emphasize science. The answers as to why this occurred, and how to reverse the trend cannot be easy, since so many bright people have struggled with the problem without real success. At the University of Chicago in the 30's, Robert Maynard Hutchins, no friend of science, still established a curriculum that devoted two full-year courses to it. The curriculum at Revelle College at La Jolla and similar ones at least have the merit of teaching genuine science, rather than survey courses, even though they don't teach very much of it. In the face of so much effort in the

---

[4] "Are Our Universities Rotten at the Core." *SCIENCE*, 236, 1165 (1987).

past of so many, any suggestions for improvement must be put in the form of topics for discussion. But even those who will disagree with my solutions (see below) may agree that we face a problem.

In the period during which the intellectual and practical content of science has expanded almost beyond recognition, the time allotted to its study has diminished, at least at Harvard, from about 33% to 6% of the total – and 6% is a bad joke.

VERTICAL LEARNING

In what way is science special? If we can analyze the difference, perhaps we can move a little closer to a solution to our problems.

Perhaps the fundamental difference is that science is more highly vertical than social sciences or the humanities (excluding foreign language).[5] In every discipline, one needs some background, but then eventually reaches a plateau where a student can branch out in any direction. Foreign languages provide examples of highly vertical disciplines; a scholar in Russian or Japanese, for example, who wishes to read the original literature in his field needs several years of study of Russian or Japanese before he or she can manage, although a student can at least obtain some understanding of the related civilizations through translations. A college student of history needs only English, and a rudimentary outline of world history; adequate texts are published in English. Science courses by contrast routinely demand many prerequisites.

An aphorism of science is that one experiment is worth a hundred expert opinions. With respect to the vertical nature of science, the course catalogs of colleges and universities provide the experiment. One does not find many prerequisites listed for the humanities. One can take a course in English history without German history, or Spanish history, or Chinese history, and conversely. But if you should wish to learn quantum mechanics, you will need to know differential equations, and to take a course in differential equations you will need elementary calculus, and to understand quantum mechanics you will

---

[5] "Deciding How Much Science is Enough," *Harvard Magazine*, May-June 1992, p. 38ff.

also need general chemistry and elementary physics, and perhaps physical chemistry as well. If you wish to study biochemistry, you will want a background of organic chemistry, and if you want to understand organic chemistry, you will need general chemistry, at least at the high school level.

These prerequisites are not arbitrary obstructions thrown in the path of students by arrogant scientists; on the contrary, they are needed background; without that background, the advanced courses are almost impossible; with it, they are quite reasonable. The various scientific disciplines require work to learn, but they are not too difficult, when taken in sequence. The vertical nature of science is unmistakable.

This statement should not be misinterpreted as a boast that science, since it is more highly vertical, is superior to the humanities, but it is certainly different. A true scholar in any field is notable. A great humanist has a breadth of knowledge that most scientists lack. In fact, the latter have often been accused of a narrowness of knowledge that many of us strive to overcome.

Learned humanists illustrate the breadth of the humanities by picking up allusions to classical and modern texts that, if understood, increase the richness of the work at hand. Perhaps I can illustrate the methodology with a low-level example. A few years ago, a movie entitled "Fried Green Tomatoes" was popular. At one point, the tom-boy heroine, at an age of perhaps ten, beautifully dressed in very feminine attire, was descending a curved staircase to make a grand entrance at an adult party. A young boy, looking up at her from the ball-room floor, sang out, "I see England, I see France," whereupon the girl, abandoning all pretense of being a young lady, stormed down the stairs and attacked the boy with her fists. Those of our generation complete in our heads the rhyme the boy started;[6] and the rhyme is necessary to understand the action in the movie. This allusion is somewhat less elegant than one to Tolstoy, or Mark Twain, or Shakespeare, and selecting it perhaps shows the uncultured mind of a scientist, but nevertheless serves to illustrate the principle of breadth.

---

[6] "I see England, I see France. I see Betty's underpants" – or Susie's or Janie's, etc. as may be appropriate.

The verticality of science, however, probably explains much of the difficulty in teaching science to non-scientists. It requires time to acquire the background that makes the learning of science reasonable, if not perhaps easy.

Further, the vertical nature of science explains why the special science courses sometimes introduced for non-scientists are such a mistake. These courses are not usually suitable as prerequisites for other courses in science. In view of the vertical nature of learning in science, courses that cannot be used as prerequisites for other courses may be wasted; this problem does not arise in the humanities, where the Core courses add breadth just as do any other courses in the area. Special courses are instituted only in the sciences. science concentrators take the same courses in Economics and Shakespeare and European History as do the humanists and social scientists.

THE PURPOSE OF COLLEGE

What is the purpose of college, anyway? Obviously, to acquaint students with the greatest intellectual achievements of mankind, but also to help them learn to live with, and appreciate, others, including those from other cultures and backgrounds. This is one reason why residential colleges are superior to commuter ones. But on the purely academic front, a major purpose of college must be to provide a student with the background for life-long learning.

One useful criterion was implied in the commencement address that Professor Roger Hildebrand[7] gave at the University of Chicago in 1971. His father, Joel Hildebrand, had been a noted professor of Chemistry at Berkeley. As a boy, he – the father – had visited Chicago to attend the Columbian Exposition of 1893, but he had ignored the fledgling University. As his son would later reflect

> It is tempting to speculate about the difference it might have made if he *had* noticed the university, and visited its classes. ...In 1893, in his own field, chemistry, my father would not have learned about

---

[7] R.H. Hildebrand, The University of Chicago *Record*, VII (9) 283 (1973).

the electronic structure of atoms and molecules because no one had discovered the electron... In physics, he would not have learned about relativity, or the uncertainty principle; in astronomy, nothing about the expanding universe. None of these concepts had been discovered...

The artists had not heard of cubism, and had never seen a Braque or Picasso... English professors were not discussing any of Hemingway, Falkner or Steinbeck, and not much of Henry James. The biologists didn't known about the genetic code. The archaeologists hadn't found King Tut. Nobody knew how to read Hittite...

Hildebrand was emphasizing that anyone who stops learning when he or she leaves college will soon be undereducated. Anyone who believes that, in his or her own field of specialization, a college education is enough is, almost by definition, describing a dead field.

The need for life-long learning however is especially acute in the sciences, where new discoveries are frequent and important. But most significantly, since sciences are so highly vertical, an individual needs a good background by the time he or she graduates from college to be able to move forward with the times. The trivial amount of science taught to non-scientists at some of our most prestigious institutions simply will not serve as adequate background. One will need to keep up with discoveries that have not yet been made. One needs a background in science now to be able to read the Tuesday science section of the New York Times, and one will need a background in science to be able to read that section in the future.

OBJECTIONS TO AN INCREASED EMPHASIS ON SCIENCE

Non-scientists at the Daedalus conference suggested that what they wanted, and all they wanted of science was a knowledge of the scientific method; they did not want much knowledge of the subject matter. But the modern world rests on the generalizations of science, the vocabulary of science, and on technology. Just the scientific method (if it could be separated from the content) won't do. No one would suggest that he or she learn French without memorizing some vocabulary; the rules of grammar aren't sufficient. Although historians have been moving away from rote memory, one still needs some sort of framework of dates in order to put any events into context.

The principles of science are based on a consideration of facts. After one has gone through the arguments that lead from facts to hypotheses and to the testing of hypotheses, one may see how science is built, but without lots of facts, and specifics for the testing of hypotheses, one doesn't understand science, much less know any.

ANYTHING WORTH DOING

On occasion, non-scientists have argued that the body of science is so vast that it cannot be learned; why try? Even practicing scientists know only a little; nuclear physicists don't usually know much molecular biology, chemists don't usually know much anatomy, and so on. Since no one knows all of science, an attempt to learn it is impractical. The same criticism could of course be made for learning of foreign language. There are so many foreign languages that an attempt to learn them all is fruitless; why, then, try to learn any? But this attitude is self-defeating, and runs counter to C.K. Chesterton's aphorism[8] that "Anything worth doing is worth doing badly." For example, if you play tennis badly, you are in a much better position to appreciate Wimbleton than if you have never played; if you play a musical instrument badly, you are in a better position to enjoy the Boston Symphony, and if you have a moderate knowledge of science, that will serve much better than the negligible input from many of today's core curricula.

---

[8] C.K. Chesterton, as quoted in the *Viking Book of Aphorisms*, W.H. Auden and Louis Kronenberger, Eds., Viking Press, N.Y. 1962.

THE UNSPOKEN ALLIANCE

The vertical nature of science is critical, but cannot provide the complete reason for the troubles in teaching it. Reams have been written about "publish or perish". Most colleges and universities try to insist that their science faculty engage in classroom teaching of undergraduates, rather than restricting their activities to instructing and collaborating with graduate students and postdoctorals. Nevertheless, in the sciences much of the faculty's emphasis is on the teaching of graduate students, rather than of undergraduates, even of science concentrators. Not all scientists truly enjoy teaching non-scientists; mathematicians sometimes openly admit that they do not even enjoy teaching physical scientists. Perhaps only 6% of the curriculum for non-scientists is assigned to science at Harvard and similar universities because of an unspoken alliance between humanists who do not wish their students to have to devote the time and effort to learn about the real world, and scientists who don't care much whether or not the unwashed are educated. But the public that sends their children to college, and have to pay for a good part of their education, should care that their children should learn some science.

MONEY

Universities are expensive. The libraries, that constitute the laboratories of the humanists and social scientists are sinks for money. Sciences require even more. Laboratories are expensive to construct and to maintain. Research requires equipment and materials. Biologists require facilities for mice and other animals, and X-ray spectrometers and computers to determine, by X-ray crystallography, the structures of complex biological materials. Chemists need half-million dollar nuclear magnetic resonance spectrometers, and expensive chemicals, which then create problems for safe disposal. The cost of the machines required by nuclear physicists is famous – or perhaps infamous. Modern science at universities would not be possible without massive Federal and other support. Outside support comes in several ways.

The National Science Foundation, the National institutes of Health, the Department of Energy and other U.S. agencies make research grants to individual university investigators for research. Those grants allow the investigators - the faculty - to support graduate students, for stipend, equipment, and supplies. The graduate students, in turn, severe as teaching assistants in the labs and discussion sections. Many fewer would be available for teaching if they were not supposed in their research, and were paid only by the university. The research Universities profit enormously, in attracting good graduate students as teaching fellows, and in financing the research that their faculty wish to carry out; no university, not even a State university, is rich enough to manage without help. Further, the universities collect huge sums in overhead, sums that effectively pay for the upkeep of the labs, and probably more. In fact, careful bookkeeping will be required to determine how much of the Universities funds are devoted to science. Incidentally, evaluations of science faculty for promotion consider not only their publications, but their success in raising research money.

WHY WE ARE DAMNED

So far, so good. But like so many other good things, the system can be abused. Many science faculty do great research, and along the way build large research groups of students and postdoctorals. The research groups require and the faculty investigator must somehow bring in huge sums to the university. Some - not all, but some - of the principal investigators who bring in the money pay scant attention to teaching undergraduates, or even refuse to do so. A rare few won't even lecture to graduate students. Many are quite unwilling to teach non-scientists. And the universities are caught. At one time, the universities and faculty were bound by an unwritten, but mutually understood, agreement. The faculty would teach both graduate and undergraduate students, and were encouraged to supervise the research of graduate students so long as the obligation to all of the students was honored. Now, because of finance, universities cannot afford to offend their research stars. The stars bring in so much money that they are in effect independent operators, and can take their prestige *and* money elsewhere if the university makes what they, the research scientists, consider excessive demands.

The psychological result when a prominent scientist refuses to teach undergraduates is devastating; that action implies that only second-class faculty give lectures. By contrast, when Enrico Fermi insisted on teaching freshman physics at the University of Chicago, his action elicited a strong competition among the senior staff in chemistry for the privilege of teaching freshman chemistry.

Could we – the educational establishment – look, to a considerable extent to colleges that do little or no research for a solution to the problem of teaching science to non-scientists? Regrettably, the faculties of these colleges are not often at the "cutting edge," and occasionally fall miserably behind the advances in science. All scholarship advances, of course, but the growth of science during the past century, as has already been explained, has been spectacular, bringing with it both excitement and difficulties.

We are damned either way. If we depend on research scientists, we will find that some - not all, but some - will shirk their obligations and get away with it. If we depend on non-research scientists, some - not all, but some - will fail to keep up with the startling discoveries that haven't yet, but presumably will be made. And the humanists, who do not seem willing to allow their students the time to learn science, are not helping. The teaching of science to non-scientists is in crisis.

WHAT TO DO?

Since so many intelligent and devoted individuals have failed to come up with a solution to our problem, it cannot be easy. but any solution to our problems must be based firmly on the realization that learning in science is highly vertical. Perhaps that realization points the way for what we should do. The universities and major colleges could demand, for admission, that all students offer evidence that they have taken and understood a high-school course in each of the three major sciences – physics, chemistry, and biology. The demand for admission to good colleges and universities is so great that they could safely enforce this requirement. Those students who could not (because their high-schools are inadequate), or who for any other reason did not fulfill this requirement would have to make it up in college, either with non-credit remedial courses or by surrendering one (or more) of their precious electives to do so. This step would go a long way to produce a scientifically literate student body, and the university professors of science probably would not object, since other people, in the high schools, would do the teaching.

Of course, this pushes the problem back a step, since high schools have difficulty finding adequate teachers of science. Besides the high schools in turn will need to find students who have been better educated in arithmetic and algebra and general science in the lower schools. Still, the desire for education at a superior college is a powerful motive, and if the colleges require better science for admission, the effect will make its way - slowly - through the entire school system.

Next, the universities could require their undergraduates to devote at least 12% of their curriculum to science. Only a few students will find this requirement intolerable. One would miss - and would want to make exceptions for - a few geniuses in ancient languages, a few spectacular artists, but most of those who would not tolerate this requirement should not in any event get the imprimatur of a degree from Princeton, Harvard, the University of Chicago, Berkeley, etc. This will not affect those colleges and universities where reasonable requirements are already in force, but will double the science required at several Ivy League universities, and at many other schools. Will humanists accede to this modest increase in science, or will they find unacceptable the diminution of students' time allotted to the studies of the humanities and social sciences from 94% to "only" 88%? Will they permit their students to enter the modern world? Will scientists agree to teach more science to non-scientists?

And what should be done with that 12% (if granted), with four half-courses instead of two? Of course, this will not make scientists of students. But, together with a better high-school background, it should make it significantly easier for the non-scientists to keep up, after college, with new developments. They would be better able to appreciate and discuss the science that hasn't yet been discovered. They would quality as educated men and women.

SCIENCE AND HUMANITIES AS COGNATE PATTERN-MAKING

D.C. PEASLEE

The following essay is an attempt to show that mechanical model-making – the hallmark of physical science – is not an anomalous outgrowth of civilization but is in fact a specialized expression of animal nervous systems operating in their fundamental mode: namely, that which has also given rise to the humanities. To distinguish the more general processes involved in humanities – and most of everyday living – we use the label "pattern-making" as being more appropriate, since these patterns are generally non-quantitative. If pattern-making is indeed the fundamental drive of the large brain of *homosapiens sapiens,* humanities and science are both artifacts of this same force and should be understood primarily in the light of that relation; comparison should then provide enhanced insight into the specific nature of each.

Because the scientific organon is specially designed for stepwise exposition, it will be convenient to develop it first. Humanities can then be discussed in terms of similarities and differences vis-a-vis science.

SCIENCE

I. PHYSICAL MODEL-MAKING

Here "physical" relates to the deterministic, quantitative aspect of science, "model-making" to its opposite: the intuitive, qualitative aspect. The basic paradigm of research in physical science can be

divided into 5 steps, suggested in Figure 1 and detailed in the
following.

1) *Observation* of repeated physical phenomena. A famous
example is the story of Galileo, who noticed the swinging of lamps in
church and wondered why those with short chains beat time faster
than those with long ones. Ultimately he deduced the pendulum
formula: $T = 2\pi\sqrt{L/g}$, where T is the time for one swing ("period"),
L is the length of the pendulum and g is a natural constant expressing
the force of gravity.

Already in this first example we see one of the characteristics of
physical science: to violate initial "gut feelings". Note that the *mass*
of the pendulum appears nowhere in the formula, contradicting the
expectation that on a given chain a heavy lamp would swing slower
than a light one. This feature makes science hard to accept: the fact
that it not infrequently contradicts naive impressions of things. If this
engenders the fear that science may destroy some of one's cherished
beliefs, the natural reaction is to turn away in the hope that what you
don't know won't hurt you – or at least won't disturb you.

Does this imply that all instinctive reactions must be abandoned
in the pursuit of science? Not at all – the second step in Figure 1 is to
invoke intuition *for* physical model-making. The most spectacularly
successful scientists are often those with the best intuitions, who can
guess where to look for answers earlier than their associates.

Perhaps one could say that gut feelings resemble pet dogs: useful
assistants if trained, burdensome to damaging if not. Moulding
intuition in a scientific discipline is rather like canine training – not
inherently difficult, but a labor of long duration. It is not for the lazy
or impatient, like the supposedly clever girl who shunned
mathematics, saying, "It really means just thinking things through,
when all I really want is answers."

2) *Modeling* consists in making some abstractions from a
particular observation and searching for similarity with facts and
models already known. There is a tendency to formulate the new
model in mathematical terms to the extent possible; for the rules of
mathematics are well established and permit precise, quantitative
predictions that can be tested.

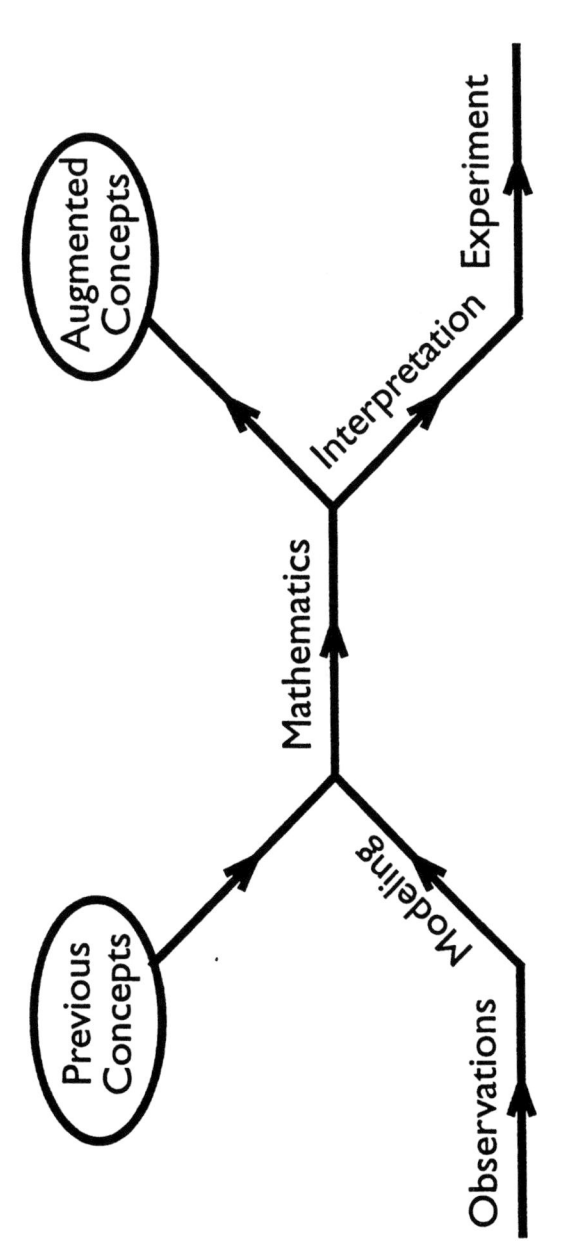

Figure I

Physical models often spring from simple mechanical analogies -
e.g., electric current in a wire can be modeled hydrodynamically as
water flowing through a pipe: the narrower the pipe, the higher the
resistance in the circuit and the higher the pressure (voltage) needed
to maintain a given current. The hydraulic model can perhaps be
extended with pressure tanks and heavy turbines to simulate
condensers and coils in the electric circuit; but at some point it
becomes more efficient to develop the electrical model in its own
terms. Nevertheless, the hydraulic model serves as a useful first stage.

To make mathematical models, intuition plays a cardinal role in
determining what the quantities x, y, z... of the formulas should
represent in the physical world. This is not a trivial question, both the
form of the mathematical expression and the physical meaning of the
variables in it being crucial to the success or failure of the model.
Engineering courses frequently include a special topic called
"dimensional analysis", which focuses on how to choose the most
appropriate combinations of variables among the physical quantities
involved. In the simplest case of a straight rod these could be the
mass M, length L, or mass per unit length M/L but not all three at
once.

3) *Reformulation* of the model means rearranging its elements so
as to predict effects not yet observed in the real world. A model that
does not allow manipulation of this sort is essentially useless to the
progress of science, however appealing it may be as a flowering of
human ingenuity. Such reformulation can be especially facile for
mathematical models – which is another, rather compelling argument
in their favor. This central step is mainly an exercise in deducive
logic. Given a model cobbled together out of immediate observation,
a fund of earlier analogies and the invariant rules of mathematics,
what does it imply? The process at this point is not unlike a detective
story on television. Given a situation (model), make a catalogue of
suspects (implications) and then pursue them (steps 4 and 5 in Figure
1).

4) *Interpretation* of the reformulated model is in principle the
reverse of step 2: that is, translation back from the abstract world of
the model to the real world of the originally observed phenomenon.
This requires more than simple reversal of the operation performed in

step 2, however. The purpose of reformulation is to provide new avenues for testing the model by comparison with the physical world. But the model itself does not usually point out where such opportunities lie, for it has merely been rearranged.

The researcher must again call on his fund of experience to estimate which newly expressed aspects of the model are feasible to test by specially designed experiments. Here is where the exercise of human ingenuity and insight are at a premium. If the model is at all complex, ambiguities of interpretation may arise, leading to more than one plausible explanation of the facts. Resolution of such dilemmas can be particularly arduous and sometimes is reminiscent of the three blind men describing to each other their first, unexpected encounter with an elephant.

5) *Testing* each interpretation is by "carefully controlled" experiment – which ideally means that all variables are held constant except the few under examination. In practice this is never perfectly achieved; but it should be possible to institute damage control by assessing the errors ascribed to uncontrollable variations. As long as these errors when cumulated remain arguably insignificant, the test can be carried through. Obviously error analysis is crucial to this process and determines the reliability of conclusions drawn from the measurement. In critical cases a major fraction of the experimental effort goes into reducing the errors.

If experiment verifies the prediction, the model gains further support and is revealed as being more general in application then just to the observation that generated it. Quite often, however, a test does not confirm the model in the new application. It is then necessary to "return to the drawing board" and try to codify or extend the model to encompass the new observation while still remaining valid for the original one. The entire cycle now repeats itself as suggested in Figure 2, where the failed test is now the observation, and the old model is included in the fund of previous knowledge contributing to the new model. Clearly, such a repetitive cycle could continue almost indefinitely, the model acquiring further generality and refinement in the process. It is perhaps this feature that prompted the philosopher of science Karl Popper to pronounce that we learn more from our failures than from our successes.

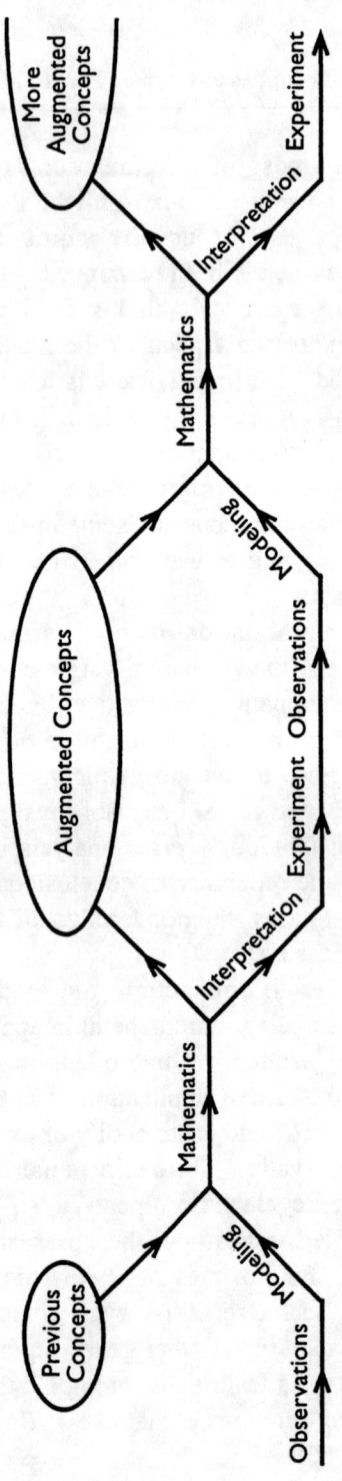

Figure 2

Summary. In this discussion the basis of scientific research is disclosed as systematic trial and error. It has two complementary strengths: the rigorous protocol described above, plus dependence at each juncture on human inspiration. Indeed, all of this is hardly more than a specialization of universal human experience; for example, Karl Popper's aphorism is predated by the general observation that "when all wrong answers have been exhausted, whatever remains – however improbable – must be the truth." Finally, the process of scientific research is fundamentally open-ended, having no fixed time scale. Once initiated, it irrevocably joins the flow of human existence.

II. CHARACTERISTICS OF SCIENCE

The success of science encourages its adaptation to other fields of human activity. Since pure science is highly specialized, such extension entails some relaxation of scientific rigor; if this can be done without sacrifice of scientific principles while at the same time providing the wider field with fresh insights, the result will be positive. Because of these constraints, not every extension will be equally effective, and there is some risk of misleading fads or even fraud developing. Up till now, however, such incidents have been quite isolated.

To consider such extensions, we may start by cataloguing some of the hard core details of scientific practice. We can then hope to follow their evolution as they impinge on wider fields. Since the archetype is physics, the details refer to that discipline.

1) *Persistence* is a necessity: repeating many cycles in Figure 2, going over and over what seems like the same old ground with only minor variations. This process can consume months and years of a physicist's life; the ultimate result is perceived as a "breakthrough", usually with insufficient credit awarded to the years of persistence behind the breakthrough.

Probably factual, the following story is illustrative. A student working toward a Ph.D. in psychology at the University of Tennessee in Knoxville aimed to elucidate personality differences among the scientists – physicists, chemists, biologists – at nearby Oak Ridge National Laboratory. On the whole no major distinctions occurred, the most striking exception being the "block test". Sets of black blocks of irregular shapes were to be fitted together to make a regular 3-dimensional figure, cube, sphere, cone, etc. – and each volunteer was timed in this effort One of the block tests was in fact impossible, although the volunteers were not so informed. Most of the participants gave up after several minutes' effort, often coming to the correct conclusion of impossibility. Here the physicists were outstanding for their persistence – they often had to be timed out. More than that, however, when the psychology student next returned to Oak Ridge, who but a physicist should accost him with the demand, "Show me those blocks again, this time I know how to assemble them." Optimism shading into arrogance is in some quarters regarded as the trademark of physicists; there is little to suggest that the physicists are much bothered by this epithet.

2) *Reproducibility* is the single most important feature of a scientific experiment and is responsible for the exhaustive publication of technical details. In principle, any one who reads the complete publication and repeats the procedure described should be able to reach the same conclusion: if not, the discrepancy must be resolved, usually by further measurement. Of course today some experiments are so large and costly that reproducing them can be prohibitively expensive. Nevertheless, Fermilab has acted on this principle and authorized 2 independent major experiments called CDF and D0 to study production of the top quark.

3) *Quantitative* results are preferred when possible. Not only do they motivate maximum precision but they readily lend themselves to error estimates. For example, we can specify a weight as 100 ± 2 grams, which means we expect it to be about twice as likely to lie between these limits as outside them. By comparison, a weight of 100 ± 8 grams is one about which we know definitely less.

Quantitative measurements not only tell us what we know but also allow us to express in this convention the limitations of our knowledge. Such self-assessment becomes less feasible, the more qualitative ("softer") the science – e.g., ecology, sociology. As a result large federal expenditures can be devoted to social programs on quite insecure premises.

4) *Limitations* inherent in scientific knowledge have already been discovered in physical problems. In the realm the very small the

invention of quantum mechanics in the 1920's led to brilliant success at interpreting atomic phenomena and initiated a revolution in 20th century physics. But quantum mechanics is a direct consequence of the hypothesis that it is *impossible* with a single measurement to know both the position q and momentum p of any object with perfect accuracy. This limitation is expressed in Heisenberg's uncertainty relation $\Delta p \Delta q = h2\pi$, where Δp and Δq are errors in the measurement. Here Planck's constant h is a fundamental constant of nature; it is so small that its presence cannot be detected with ordinary macroscopic objects, but it becomes overwhelmingly important for small ones like electrons in atoms.

At the other extreme of large-scale phenomena, the theorem of MIT meteorologist E.N. Lorenz is gaining wide acceptance: namely, that weather patterns in the atmosphere are essentially chaotic. This means again that prediction is impossible beyond a short term like a couple of days or at most weeks. No matter how large our computers nor how much memory of past weather patterns they maintain, precise details beyond that short time frame can never be predicted reliably.

If even physics – the most precise and quantitative science – has inherent limitations, it is reasonable to expect that extensions to further fields must be even less deterministic.

5) *Generalization* is the main technique used in the advancement of physical models: that is, later refinements encompass more phenomena but usually do not discard the former models, instead retaining them as special cases. The classic example is Einstein's special relativity, where Newton's mechanics are not invalidated but defined as applicable only in the most common case, where all (relative) velocities are small compared with that of light, c=1 foot/nanosecond, where 1 ns.=10^{-9} sec. =1/1,000,000,000 sec.

A striking recent example of generalization in physics has been the emergence of the electroweak force. During most of the 20th century it has been conventional to distinguish 4 basic forces in nature: gravitational, electrical, radioactive or "weak nuclear", and nuclear. The nuclear forces are much the strongest and gravitation the weakest; in between these, radioactivity appeared to merit the appelation "weak" by being much less powerful than electricity. Over the past decade or two, however, it has been established that electricity and radioactivity are two expressions of a single basic force, now called "electroweak." From one point of view this is a monumental generalization: the number of basic forces in the universe being reduced by 25%.

Incidentally, physicists beginning with Einstein have been searching for a "unified field theory" to derive all observed forces from a single origin. Even the generalization just described, however, brings with it no assurance that such a theory (e.g., currently popular "string theory") would correspond to the real world.

6) *Cumulative* knowledge can be cited as a feature of scientific research. Later models are built on earlier ones without entirely discarding them, so the corpus of knowledge continues to grow. Science is a conservative discipline.

Actually, in a sense all knowledge is cumulative; but an individual acquires the everyday knowledge required to live in his milieu over decades in his early life by means that are often informal. Because science is an artificial discipline designed for special purposes, the cumulative process becomes much more evident. It also proceeds at a forced pace compared with the more gradual accumulation of non-scientific knowledge – but this is more a circumstance than a fundamental difference. Whether that forced pace is a necessary concomitant of science is not clear, however; it may be ultimately counterproductive, turning people away from the subject.

III. EXPANSION INTO NEIGHBORING FIELDS

A major growth area for quantitative science has for some time been at its periphery into tanget fields. A common instance is where an originally descriptive science – geology, biology, in some respects astronomy – becomes more quantitative and changes its name accordingly to geophysics, biophysics, astrophysics. These sciences are somewhat less pure in the sense that direct experimentation is less feasible, especially in astrophysics but often in geophysics. In such cases the exercise of inventiveness is to set up laboratory experiments designed to simulate conditions observed in the external world – and then observe whether developments in the laboratory mimic those in nature. In this procedure quantitative measurements are of great importance.

Upon further extension to the social sciences, not only quantitative measurement but even the more primary concept of *measurability* must be introduced. In economics, for example, the one-time notion of "economic man" has fallen into disuse because it did not lead to any unequivocal, measurable consequences. This remains a central problem in social sciences: namely, to identify significant, reproducibly measurable parameters.

Can science extend even into the classical humanities? Probably to some degree: surely the present explosion of telecommunications will lead to further scientific analysis of communication in general and language itself. Neuroscience may ultimately (as neurophysics?) succeed to quantitative models for communication, cognition, memory, possibly even some emotions.

IV. SCIENCE IN EDUCATION

By its nature – with ever-ramifying models – the scientific approach can be expected to continue invading traditionally non-scientific areas. This should supply motivation for every modern citizen to have some appreciation of how science is done, even if he or she is not a specialist. In the educational curriculum science would hopefully receive equal emphasis with other subjects. Before this golden age arrives, however, it may well be necessary to develop "remedial science" courses at senior high school/college levels to compensate students for insufficient mastery in lower grades. Such courses would have to emphasize the difference between unfamiliarity and intrinsic difficulty, by filling in background that is lacking and insisting that science is little more than codified common sense plus some elementary math.

Perhaps hardest of all today will be to persuade students that hard work is not permanently injurious. One of the most sacrosanct principles in physics is *unitarity*, which translates directly into the statement that "what you get out (of anything) is just equal to what you put into it."

HUMANITIES

I. ANIMAL PATTERN-MAKING

Consider the newborn animal with few if any pre-imprinted patterns but enough sense organs to enable adaptation to its environment. How does the newborn come to "know" this environment? We can answer by again resorting to Figures 1 and 2, illustrating a remark attributed to Piaget that early childhood learning is a paradigm of science. Now in the figures the lowest horizontal line represents sense inputs, which are essentially random. The middle line is where a repeated pattern is recognized and selected – with aid from relevant memories, if any; the developed pattern is then also stored in the memory which is now the top line.

What inborn mechanisms perform the pattern recognition on the middle line of the diagrams? Whether or not there are others, a temporal sense of before/simultaneous/after is probably a principal one. The corresponding pattern is one of simple *association* of otherwise independent inputs: in a human child, the warmth of the parent's touch and the sound of the parent's voice, for example. Such association becomes more complex with practice but always preserves the features of simplification and consolidation; effects first perceived to be separate are understood as parts of a larger whole. Somehow this is a relief to the memory and the whole mental structure, resulting in a sense of enlightenment, sometimes called the Aha! syndrome. This internally generated sense of pleasure provides a strong impetus to further pattern-making.

The primitive inputs at this stage all derive from the five sense. For humans their strict order of importance is sight, sound, touch, smell and taste; for lower animals they are more equal and even of variable priority. In all higher animals another sense develops early which we may call kinetic, being the feedback from motor activity. It is a sense of moving the muscular apparatus of the body, independent of sight or other inputs. Its existence becomes clear in a pathological case where motion is felt in an amputated limb through nerve patterns laid down long earlier.

Arising from the sense of motor activity is the pattern of motor control. Development of such patterns is encouraged by rattles and toys in the baby's crib or is likewise seen in the a rather purposeful play of kittens. Kinetic patterns are especially significant, for they introduce the concept of *control*. Pattern formation now proceeds more rapidly.

In terms of Figures 1 and 2, motor control is the first element analogous to the diagonal line from the central to the lowest horizontal lines. The infant makes a controlled experiment – push the ball and it rolls, push the piano and it doesn't. This is the infant's first scientific experiment – done spontaneously before walking, talking or even understanding parents' instructions. How does this instinct get so lost in later years?

There is evidence attesting to the importance of primitive pattern-making as a necessary life function. Sensory deprivation experiments involve suspension of mature adults in a liquid medium with no inputs of light or sound and minimal tactile (temperature) sensations. The common result is the onset of hallucinations: that is, the nervous system needs continuous input and if cut off from any external supply regurgitates random patterns from memory. The best defense against this effect is to concentrate attention on higher patterns already stored in memory - i.e., to provide self-generated input in a controlled rather than uncontrolled fashion.

II. SECONDARY HUMAN STAGE: TALKING, WALKING

Verbal pattern-making is peculiarly human, at least in degree, and constitutes the greatest single advance in the individual's development.

The baby is provided with every incentive to verbalization, and imitative language efforts begin very early but in the first year are impeded by a lack of motor control. Although language patterns are not inborn, there is a suspicion that human brain development in the first decade is programmed to facilitate language acquisition. Young children are observed having little difficulty in picking up two languages that for adults are totally distinct: e.g., the children in a Finnish-American family speaking both equally well and knowing instinctively which family members respond to each language. This choice of separate vocabularies is at first made without explicit recognition of different languages.

The tremendous impetus provided by language is that now the individual can receive patterns, preformed, without having to develop them slowly by personal observation. The whole knowledge of the community becomes in principle accessible. In an advanced society this is likely to be more than an individual can encompass in a lifetime, so a selection must be made from an overabundance of riches. Accordingly, the educational system of the community becomes a matter for continuous debate.

The verbal facility to input whole patterns is a great stimulus to formation of higher-order patterns: namely, more complex ones using as basic elements not so much pure sense data as other patterns already laid down in the memory or imported through verbal channels.

At about the same time as verbal communication is initiated, the human infant learns to imitate the bipedal locomotion displayed by all older humans. This now may be regarded as a major undertaking which engages a substantial fraction of the human forebrain's capacity. The motor control, like verbal facility, continues to improve with practice throughout the maturation years.

These two great advantages of *homo sapiens sapiens* do not come without cost. Untrammeled elaboration of mental patterns – especially in verbal, qualitative form, can lead to concepts having little resonance in the real world. Perhaps worse, the exercise of motor control has always led to control of the physical environment and thence soon to the notion of controlling other human beings. Today the development of mechanical facilities and mass communication have brought us to the point where our reach greatly exceeds our grasp. Gargantuan, centralized social experiments are based on concepts pleasing to a few individuals' imaginations but having tenuous connection to fundamentals of human nature. The result has been to characterize the 20th century as the first in which murder of individuals by their own governments has exceeded foreign wars as the great decimator of humanity.

III. MATURE DEVELOPMENT

The discussion above suggests that pattern-making is the primary inherent drive of human mental endowment. This activity provides a certain kind of self-generated pleasure; like happiness, the pleasure is transistory and leaves the individual seeking for more. With this incentive we can expect pattern-making to continue throughout the life span of the individual. Evidence for this is the ceaseless fascination with useless novelty – from fashion in clothes to organized sports to TV news broadcast to exotic travel for senior citizens.

A positive feature of continual pattern-making is the tendency to melioration. The pleasure is greatest when a superior pattern is constructed out of two lesser ones: not only because of greater efficiency in memory storage but also the sense that something constructive – call it greater insight – has been achieved. With this insight may come opportunity to apply the new pattern to further improvements in understanding.

Not a big extrapolation from this discussion is to characterize human beings as compulsive explainers. If a gap in some pattern exists because of insufficient knowledge, the instinctive drive is to fill it with imagination – i.e., patterns constructed entirely from stored neural models – and thereby form a pleasingly complete if not entirely realistic pattern. This accords with Schumpeter's interpretation of myth among primitive people. These were not fairy stories told for amusement but serious efforts to explain natural phenomenon not comprehended by their limited understanding of the physical world.

Now the western world is at the opposite extreme, where the fervor for abstract pattern-making has tended to divorce us from the real world. The consequences range from ludicrous to disastrous – cf. Thomas Sowell, *Is Reality Optional?* (Hoover Institution Press, Stanford, 1993).

IV. DEVELOPMENT OF SOCIETIES

According to the felicitous phrase, "ontogeny recapitulates philogeny", the history of a society mirrors the life cycle of the individual. In this way one can understand the dominance of humanities in all previous cultures, up to and including the educational curriculum in ours.

The coming together of tribal, frequently nomadic groups to form settled communities appears as a great advance, comparable to the major development of talking and walking in the individual. It would involve considerable negotiation and could hardly have predated the full flowering of human speech. Which tribal language predominated would depend on personality and forcefulness among the founding chieftains. The kinetic element – walking in the case of the individual – is reflected in the use of animal power (no doubt frequently human slaves captured in raids) for construction projects in the community: roads, waterways, fortifications. Leaders first rose to power by personal prowess, like bullies of the pre-adolescent playground.

As the settlements became larger and more complex, explicit rules were needed to maintain domestic order. Among the earliest were moral codes, aimed primarily at teaching the individual how to live non-destructively in a closed community. Later came laws like those of Hammurabi, where the community itself would intervene to provide protection and redress for the citizen against injury by his neighbor. All these developments depended strongly on verbal constructions and qualitative patterns.

Quantitative ideas were at first limited to two opposite extremes: practical measurements of land areas, leading to the development of geometry; and astronomical measurements by the priestly class, who were concerned with making dependable predictions related to cycles of the seasons and who developed numerical methods of calculating these cycles. Perhaps the individual analogy is to the adolescent flourishing of interest in sports depending on distance, direction, accuracy and in observing and forecasting the weather.

When leisure became available for speculative thought, it was natural that philosophy should focus almost entirely on what are now called the humanities; geometry and astronomy were enshrouded in an aura of magic and mystery. This hegemony has persisted through all known societies until the present, the post-renaissance civilization of western Europe being the first to incorporate science as one of its major pillars. Once this breakthrough did occur, its advantages spread it worldwide; the fascinating question is why it happened first in Western civilization, given that other advanced societies might have produced it.

V. SCIENCE AND MORALITY

There is a popular myth to the effect that science is antithetic to morality, that its goal is "value-free knowledge". This becomes an arraignment only by addition of the unspoken assumption that science proclaims itself as the arbiter of all knowledge. On the contrary, an important aspect of science is to define accurately the limits of any investigation. These limits are narrowly associated with the scientific techniques being employed. The bulk of our accumulated knowledge still lies outside science.

On the other hand, it is revealing to note that the successful *practice* of science displays most of the primary virtues of the western world. The list below is extracted from William J. Bennett's *Book of Virtues* (Simon and Schuster, New York, 1993) and Armstrong Williams' column of February 11, 1996, in the Boston *Globe*.

1. Perseverance: determination: Figure 2 *a fortiori*.
2. Self-discipline: mastery of difficult techniques, mental and physical; maintenance of competence a personal responsibility.
3. Honesty: required to an exceptional degree, for science could not progress otherwise; on a practical level, dishonesty strongly discouraged because so easily exposed.
4. Faith: emphasized by physical chemist H.S. Taylor, sometime Dean of the Graduate School at Princeton, who liked to pint out that the most basic principles of science must be taken on faith in much the same way as the tenets of a religion. Conservation of energy in every reaction is a paramount principle of science – but who has observed every reaction in the universe? All we can say is that we have never seen it violated in any reproducible study. The generalization is akin to a religious leap of faith.

5. Hard work: generally cited as a virtue, Bennett thinks it must also
    be a rewarding activity in itself. Certainly scientists become
    highly absorbed in their pursuits, perhaps because they are testing
    their mettle against the challenge of a formidable but fair
    adversary. Nature may dissemble its secrets craftily, but it never
    cheats by altering the rules.
6. Hope, optimism: engendered by the faith that nature plays a fair
    game. In Einstein's words, "God does not play dice with the
    universe".
7. Golden rule: the peer review, which resembles democracy is
    having an excess of faults until compared with any alternative.

Does such a listing – which could be extended – imply that
science should be promoted to teach morality? Probably not; it comes
too late in human development. Here we can reverse parallels drawn
in the preceding section. In community development moral values – a
qualitative concept – necessarily mature much earlier than an
articulate science. Likewise, an individual's moral outlook must be
inculated early, probably during the pre-adolescent stage of language
development. At a later stage science can confirm and reinforce these
moral values, but the basis must be already laid down.

This discussion leads to a couple of suggestions for university
teaching. The importance of morality, not only in general but as a
basis for the pursuit of science, suggests a revival of the centuries-old
requirement of a freshman course in moral philosophy. Dare one
propose calling it "Remedial Morality 101"? The other suggestion is
that science faculties interested in redressing the balance between
science and humanities should recognize that humanities, especially
language and verbal images, represent practically the only channel of
communication for reaching students with weak backgrounds in
science.

Aristole, according to Bennett (op. cit), considered that moral
education is based on precept, habit and example. Precept must come
first, but science is good training in *habit*; and *examples* of character
are legion among the lives of great scientists. Even Einstein
considered character more significant than brilliance in a scientist.

INDEX

A

abstraction of mathematics, 45
acetylcholine, 22, 23, 25, 26, 28, 32,
  35, 38
Adults, 1, 12
amyloid plaques, 21
association, 96

B

Babies, 2, 3, 5, 6, 12
behavior, 5, 6, 8, 9, 10, 12
brain disease, 15

C

Channel Coding Theorem, 47
Characteristics of Science, 91
childlike, 10
Cholinesterase inhibitors, 25
Cognate Pattern-Making, 85
cognition, 3, 4, 14, 28, 32, 95
college, 71, 75, 77, 78, 80, 83, 95
communication, 20, 43, 44, 46, 47,
  48, 95, 98, 102
compact disk systems, 42
computation, 46
computer-aided drug design, 32, 35
Construction of Reality, 2
control, 89, 96, 97, 98
Core Curricula, 73
cortex, 15, 32, 55
culture, 1, 9, 41, 43

D

dementia, 15, 21, 23, 26
digital computers, 48
Digital Information Age, 41, 49, 51
digital video disk, 42
direct broadcast satellite, 42
discrete information, 42

E

elderly, 15, 21
electroweak force, 93

F

Freud, 1, 7, 8, 9, 10, 11, 12, 13, 14

G

Generalization, 93
gut feelings, 86

H

hearing, 4, 8
high definition television
  broadcasting, 42
hippocampus, 15, 20, 22
humanities, 1, 71, 75, 76, 77, 83, 85,
  95, 100, 102

I

infancy, 4, 5, 14
information representation, 45
integrated circuits, 43, 48, 50
intelligence, 5, 64
Interpretation, 88
intuition, 56, 86, 88

K

Kant, 2, 3, 12, 14
Kilby, 41, 42, 43, 49

L

life-long learning, 77, 78
linopiridine, 25